# Zero Fault

# Table of Contents

# Zero Fault

## By
### Brett Shayler

# Dedication

I dedicate this book to every young dreamer who enjoys showing Dressage horses and dares to chase their passions, big or small, and to children who seek comfort and support in the steadfast friendship of an animal companion. This connection goes beyond language and nurtures a lifelong affection. This story celebrates those who possess the quiet determination necessary to overcome obstacles, persevere through setbacks, and believe in their potential, all while being supported by a loyal companion.

It is for the children who spend countless hours honing their skills, whether mastering a challenging equestrian feat, perfecting a musical piece, or building an intricate Lego castle—this dedication is for those who value initial commitment. During mornings and late nights, hard work and determination help children achieve their goals, understanding that the journey is just as crucial, if not more important, than the ultimate success. It's a journey made easier and more rewarding with the unwavering support of friends, family, and animal companions, a reminder that dedication and support are key to success.

This book also celebrates the incredible bond between humans and horses—a relationship built on trust, respect, understanding, and a shared love for the beauty and power of the natural world. It acknowledges the patience, care, and consistent attention that went into developing this significant connection. This tale resonates with anyone who has felt the magic of horse connection, shared peaceful moments, or enjoyed thrilling rides. May the following pages inspire you to reach for your dreams, nurture your friendships, and cherish the bonds that enrich your life. Diligence, commitment, and self-belief yield remarkable results.

This is the third book in my equestrian series about performance horses and their riders. Each book will delve deeper into the unique challenges and triumphs that young equestrians and their loyal companions face. From the thrill of competition to the quiet moments of connection, these stories will celebrate the unbreakable bond between horse and rider, highlighting the lessons learned, and the friendships forged along the way. This book is for parents, guardians, and mentors who support and inspire young individuals' passion for showing horses. Your unwavering belief, patience, and dedication to their growth contribute significantly to their success. For further details regarding the trainers, the event, and any associated show information, we invite you to explore the provided website. The association's website, which is located at www.ushja.org, is provided here for your convenience. We want to express our heartfelt thanks for the ways in which you guided them, offered them your support, and inspired them along the way. The impact of your love and support was substantial, leading to a significant improvement in their journey.

# Chapter 1: Legacy's Long Shadow

The scent of hay and leather hung heavy in the air, a familiar comfort that did little to soothe the knot of anxiety tightening in Leah's stomach. The sprawling stables of Blackwood Manor, nestled amongst the rolling Kentucky hills, were a monument to her family's equestrian legacy—a legacy that felt more like a lead weight than a source of pride. Sunlight streamed through the tall windows, illuminating the polished brass tacks and gleaming trophies lining the walls, each a testament to her older siblings' achievements, which loomed over her like dark clouds. She traced the intricate carvings on the antique saddle rack, the cool wood starkly contrasting the heat rising in her cheeks.

Every morning began the same: the rhythmic clip-clop of hooves on the gravel driveway, the hushed whispers of grooms preparing the horses, and the ever-present pressure to excel. It wasn't the championships looming—the regional championships, a pivotal stepping stone towards national competitions, the culmination of years of training—but the constant, subtle, and sometimes not-so-subtle comparisons to her older brothers and sister. They were all naturally gifted riders, champions, their names whispered with reverence in equestrian circles. And Leah, despite her talent, often felt like she was perpetually playing catch-up.

Her father, a former champion show jumper himself, meant well. Though often laced with a critical edge, his advice stemmed from a deep-seated desire to see her succeed. Yet, his well-intentioned critiques usually felt like sharp barbs, highlighting her weaknesses rather than celebrating her progress. "Your approach to the obstacle needs more precision, Leah," he'd said, his voice carrying the weight of expectation. Or, "Your posture needs correcting; your sister had that mastered by your age." His words, intended to guide, often left her feeling inadequate.

She carried the weight of her family's reputation on her shoulders, an almost unbearable burden. The Blackwood name was synonymous with excellence, a standard that seemed impossible to attain. She frequently looked at her reflection in the polished surfaces of the tack room, seeking the qualities that appeared to be inherent in her siblings. She saw potential—a raw talent pulsing beneath the surface—but self-doubt shrouded it, persistently whispering of all her unachieved goals.

The regional championships weren't merely a competition; they were a referendum on her abilities, a chance to prove to herself, as much as anyone else, that she deserved her place among the Blackwood riders. Usually, finding comfort and solace in her horses, she noticed even they seemed to pick up on the anxiety that she was feeling. She dedicated additional time to grooming and caring for her chestnut mare, a fiery beauty she affectionately nicknamed Whisper. However, the mare's registered name was the more formal Storm's lil Whisper. As she gently brushed Whisper's coat and murmured reassurances, a moment of calm washed over them, temporarily escaping the immense pressure. Whisper's soft breaths and warm trust grounded Leah in the present, a small anchor in the storm brewing inside her.

Blackwood's stable: a training ground reflecting family life's energy and stress. The scent of liniment and sawdust mingled with the nervous energy that hummed in the air as riders prepared their mounts. Leah overheard her father discussing strategies with her trainer, Mr. Henderson, a seasoned equestrian coach with a gruff exterior and a heart of gold. She could perceive the growing anticipation, the unspoken expectations weighing heavily in the atmosphere, creating a tangible sense of pressure that heightened her concerns. The grandfather clock's rhythmic ticking mocked her; every tick marked her proximity to the competition, to the decisive moment.

She found solace in the rhythmic thudding of her heartbeat, trying to match it to Whisper's steady breathing. Each breath reminded her to center herself and draw strength from within. She felt Whisper's trust in her, a bond forged through countless hours of training, silent communication that transcended words. Yet, the looming shadow of her family's legacy continued to cast its influence over her, shaping every thought, every movement, every breath.

She spent hours cleaning Whisper's tack, polishing the bridle until it gleamed, and checking the girth and stirrups. It was a ritual, a way of calming her nerves, of finding a semblance of order in the chaos swirling within her. Each meticulously executed task provided a momentary sense of control, a counterpoint to the feeling of being swept along by the tide of expectation.

As if the days weren't bad enough, the nights presented an even greater and more intense suffering. The pressure on her chest was so heavy, as if someone were sitting on her, that she would wake up in a cold sweat from the sheer force of it. Even the seemingly insignificant aspects of the day, such as selecting her riding attire and meticulously choosing a hairstyle, felt incredibly overwhelming, with each small decision adding to her immense pressure. The usual thrill of ascending was utterly overwhelmed and consumed by a vast, overwhelming sea of doubt and anxiety.

The pressure to compete wasn't the source of the burden. She felt desperately unprepared to uphold the legacy, a legacy that demanded nothing short of excellence and perfection. Blackwood's name, though renowned, proved to be a double-edged sword, capable of both elevating and jeopardizing his reputation and prospects. But, with the championships looming ever closer, a quiet and determined defiance grew within her. Despite the weight of her family history, she would bravely confront the challenges ahead, refusing to allow her past to dictate her future. Having her loyal companion, Whisper by her side, she was determined to forge her path according to her desires and beliefs. She was planning to ride her horse soon and compete in the upcoming competition. At long last, the moment arrived when she would uncover and fully understand her hidden talents and abilities, allowing her to reach her true potential.

# Chapter 2: Meeting Emma the Rival

The air crackled with a unique energy that morning. A sharper, more confident rhythm punctuated the usual nervous hum of the Blackwood stables. It emanated from a tall, slender girl with sun-kissed skin and a cascade of blonde hair pulled back in a sleek ponytail. She moved with an effortless grace bordering on arrogance, her movements precise and economical as she guided her powerful Hanoverian horse through intricate dressage exercises. The stallion responded to her every subtle cue with an almost preternatural obedience, its movements fluid and powerful. Leah watched, mesmerized and simultaneously irritated, from across the arena.

People whispered about Emma, the competitor whose name inspired awe and apprehension. Leah had heard the stories, the tales of her flawless victories, her almost supernatural ability to connect with her horses. Now, as Leah laid eyes on Emma, she felt a surge of something resembling fear. Not of Emma, necessarily, but of the effortless confidence she radiated, a confidence that Leah desperately craved.

Emma's movements contrasted with Leah's, which were often hesitant and guarded, reflecting the internal battle within her. Emma was fluid, bold, and almost defiant. She challenged the horse, course, and the world. Her posture hinted at something unsettling, a subtle haughtiness pervading her. Leah unconsciously tightened her grip on Whisper's reins, a nervous habit she hadn't realized she'd developed. Sensing her rider's unease, Whisper nudged her gently, her soft breath calming the tense atmosphere. Leah took a deep breath, trying to calm the flutter of apprehension in her chest. She reminded herself that she had spent years honing her skills and possessed the talent, determination, and partnership she needed to compete with Whisper.

As Emma completed her routine, a small crowd of onlookers—fellow riders, trainers, and even a few spectators who had arrived early—applauded.

Emma acknowledged their praise with a curt nod, her expression remaining impassive, almost disdainful. It was a subtle gesture, nearly imperceptible, but it spoke volumes about her personality. It was a deliberate, calculated display of self-assuredness, a quiet proclamation of dominance that sent a shiver down Leah's spine.

Mr. Henderson, Leah's trainer, approached her, his usually gruff demeanor softened with a hint of concern. "She's something, isn't she?" he said, gesturing towards Emma, who was now dismounting, her movements as graceful as those of her horse. "A natural talent. But don't let that intimidate you, Leah. You have your strengths."

"She's...different," Leah mumbled, unable to articulate precisely what unsettled her about Emma.

"Different, yes," Mr. Henderson agreed, his eyes twinkling. But remember, show jumping isn't about skill but about heart. And I've seen plenty of heart in you, Leah—more than you give yourself credit for." His words provided comfort, reminding her she wasn't facing this situation alone.

Leah watched as Emma effortlessly saddled her horse, her movements as precise and efficient as those of a seasoned professional. There was no wasted motion, hesitation, or visible sign of nervousness. Even how she adjusted the saddle and checked the girth spoke of a deep understanding of her mount.

Emma and Leah's eyes met for a moment. A ghost of a smile flickered across Emma's lips, but it held no warmth. Leah felt her cheeks flush, a mixture of embarrassment and annoyance. She spun away; her gaze settling on Whisper, patiently waiting. Her calm demeanor was a comforting contrast to the unsettling encounter with Emma.

The days leading up to the championships were a blur of intense training sessions, meticulous preparations, and the ever-present shadow of Emma's seemingly effortless prowess. Leah constantly compared herself to Emma, scrutinized her performance, and highlighted her weaknesses. She practiced meticulously, pushing herself harder than ever, but a nagging voice of doubt still echoed in her mind. It whispered insidious comparisons: Emma would never struggle with that; Emma would have mastered that by now.

She sought advice from her father, but his words, while well-intentioned, only amplified her anxiety. "You need more aggressive lines, Leah," he said, his usual critical tone more pronounced than ever. "Emma's precision is something you could learn from." His words, intended to be constructive, felt like a slap in the face, a stark reminder of the gap between her performance and Emma's apparent perfection.

The pressure mounted, and the weight of expectation became almost unbearable. She retreated into her thoughts, replaying Emma's movements in her mind, dissecting her technique, and trying to understand the source of her seemingly effortless skill. The more she analyzed, the more insecure she became. The differences between their styles, approaches, and personalities seemed insurmountable.

She sought solace in the long hours spent with Whisper, grooming her coat until it shone and whispering words of encouragement into her soft ears. Whisper's trust and unwavering confidence in Leah provided a small anchor in the storm of self-doubt. She knew she had to find strength, overcome insecurities, and focus on her capabilities.

As the competition drew nearer, Leah realized that her rivalry with Emma was not about winning or losing. It was about confronting her self-doubt, accepting her strengths and weaknesses, and finding the courage to believe in herself, regardless of Emma's overwhelming confidence and apparent perfection. Competing against Emma, overcoming her internal struggles, and finding her voice amidst her family's legacy was the challenge. Usually comforting, the scent of hay and leather now held a sharp edge of anticipation. Gaming begun; the stakes were unprecedented.

# Chapter 3: Preparing for the Big Day

The days melted into a blur of sweat, exertion, and the ever-present scent of hay and leather. Nervous energy filled the usually calm Blackwood stables. Each rider, each horse, seemed to vibrate with a barely contained power, a potent cocktail of anticipation and anxiety. The air crackled with unspoken rivalry, the subtle tension between ambition and fear palpable in every movement, every whispered conversation.

Leah's training regime became brutal, a relentless push to refine every aspect of her performance. Mr. Henderson, usually a man of measured words, pushed her harder than ever, demanding precision and perfection. He drilled her on her approach to the jumps, the angle of her turns, the subtle shifts in her weight, and the almost imperceptible adjustments of her reins. He scrutinized every detail, dissected every flaw, and meticulously celebrated every improvement.

The physical demands were immense. She spent hours in the saddle, her muscles screaming in protest, her body aching with fatigue. Yet, she pushed on, driven by a fierce determination to prove herself, not to others, but to herself. She learned to listen to Whisper's subtle cues, anticipate her every movement, and feel the rhythm of her breath and the tension in her muscles. Their partnership went beyond the physical; it was a silent conversation, and a shared understanding that extended beyond the arena.

The mental demands were even more grueling. The pressure to perform, the weight of expectation, pressed down on her like a physical weight. Every jump was an internal battle, a constant struggle against self-doubt. The ghost of Emma's effortless grace haunted her, a constant reminder of her own perceived shortcomings. She replayed Emma's performances in her mind, analyzed her technique, and searched for the secret to her seemingly effortless skill. It was a fruitless exercise, one that only amplified her insecurities.

She spent evenings studying videos of past competitions, analyzing other riders' techniques and strengths and weaknesses. Meticulous planning allowed her to strategize each jump, visualize her approach, and mentally rehearse her movements until they became second nature.

All the riders, who seemed to share Leah's unease, felt the palpable tension. They pushed themselves harder, their training sessions becoming more intense, their competitive spirit burning brighter. A focused intensity replaced the usually jovial atmosphere of the stables, a silent battle waged with sweat and exertion. Quiet determination and a shared understanding of the monumental task ahead replaced the familiar banter between riders.

One evening, Leah sought her father, hoping for reassurance or encouragement. She found him in the tack room, meticulously cleaning Whisper's tack, his movements as precise and methodical as always. He looked up as she approached, his eyes, usually twinkling with amusement, clouded with concern.

"It's getting close, isn't it?" he said, his voice softer than usual. Leah nodded, unable to speak, the lump in her throat constricting her words. He put a hand on her shoulder, a gesture of unexpected tenderness. "Don't let the pressure get to you, Leah. Remember why you started. Remember the joy of riding, the thrill of competition."

His words, though simple, felt like a lifeline and reminded her of the pure joy that had first drawn her to equestrian sports. Winning no longer mattered; hearing him brought back her passion, her dedication, and her skill.

The last days before the competition were a blur of meticulous preparation. We thoroughly checked and double-checked every piece of equipment: the saddle, the bridle, the stirrups, the boots, and even the most minor details. They groomed Whisper's coat to a flawless shine and meticulously cleaned and polished her hooves. They left no detail to chance. It was a testament to their shared dedication and silent understanding.

Leah spent hours with Whisper, grooming her, talking to her, whispering words of encouragement and reassurance. Whisper's gentle nudges and calm presence comforted Leah, easing her anxieties. They shared a connection, a bond of trust and mutual respect that extended beyond the physical. Their shared commitment was a silent pledge to face the challenges ahead together. Competition even stables thrummed excitement, nervousness. Nervous energy filled the air. The riders showed focused precision and efficient movements. The horses mirrored their riders' anxiety, their movements revealing the rising tension.

Leah spent the evening alone with Whisper, checking her tack one last time with practiced ease. She ran her fingers through Whisper's soft mane, whispering words of encouragement into her ears. Night's quiet brought her calm, accepting of upcoming challenges. She knew she was ready. She knew she and Whisper were a team, bound by a shared love for the sport and a mutual commitment to give their best. Expectation would be immense tomorrow, but she would face it, not alone but with her trusty steed. In the stable, the only sound was the gentle, steady rhythm of Whisper's breathing, a calming presence that cut through the storm of anticipation—after weeks of preparation and anticipation, the momentous day had finally dawned.

# Chapter 4: Nerves of Steel or Shattered Dreams

As the morning broke, the air thrummed with potent energy, a thrilling yet terrifying force that filled the atmosphere. A vibrant atmosphere filled the showgrounds, a kaleidoscope of color and a symphony of sounds creating a mesmerizing buzz. A crisp breeze caused the banners to flap energetically. The sun glinted off the glossy coats of the horses, and the excited chatter of the many spectators filled the air with a vibrant hum. Leah found the lively atmosphere oppressive rather than festive; the tension escalated.

For weeks she had carried the crushing weight of expectation, and now it pressed on her like a physical burden, suffocating. She watched the other riders prepare, their movements fluid and confident. Some meticulously checked their tack, while others calmly groomed their horses and engaged in lighthearted banter with their friends. But Leah could only focus on the tightening knot in her stomach. Her carefully planned routine seemed to vanish from her memory, replaced by a relentless tide of self-doubt.

Whisper, usually a pillar of calm amidst the chaos, seemed to sense her rider's anxiety. She shifted restlessly, her usually calm demeanor replaced by a nervous energy mirrored in Leah's own. Leah groomed the extra whisper, brushing her coat until it shone, whispering words of reassurance in her ear, trying to instill the calm she desperately needed. But the words felt hollow, even to her ears. The calm didn't come.

Then she saw Emma. Emma, radiating confidence, moved with an effortless grace that seemed to mock Leah's internal turmoil. She executed her warm-up flawlessly, and each jump showed precision and style. Emma's horse, a magnificent chestnut stallion, responded to her every command with effortless obedience. They were a picture of perfect harmony, starkly contrasting Leah's feelings of apprehension and inadequacy.

The sight of Emma's effortless performance only amplified Leah's anxieties. Doubt, a venomous serpent coiled in her heart, whispers insidiously in her ear. What if I fail? It hissed. What if I let everyone down? What if I'm not good enough? The questions spiraled, each feeding the next, until Leah felt overwhelmed by a wave of despair. She tried to focus on her breathing, on the rhythm of Whisper's breath against her leg, but the images of Emma's flawless performance kept intruding, clouding her thoughts and undermining her confidence. She closed her eyes, trying to visualize her course and rehearse her movements in her mind, but the images were blurry and indistinct, overshadowed by the growing sense of dread.

Her father approached, his eyes filled with quiet concern, mirroring her feelings. He placed a reassuring hand on her shoulder, his touch a silent message of support. "Remember what we talked about," he whispered, his voice a calming and steady force on her frazzled nerves. "Focus on your riding, on your connection with Whisper. The rest will follow."

His words revived memories of what first attracted her: the ride's exhilaration and the connection with her horse. She took a deep breath, trying to push aside the doubts that plagued her and focus on the task at hand. Winning was secondary to partnership, trust, and shared journey. She looked at Whisper, her eyes meeting her horse's intelligent, reassuring gaze. Whisper nudged her gently, her warm breath a silent promise of support.

Course tension felt intense. Crowd energy, rhythmic hoofbeats, and pre-jump tension filled the air. Leah felt the familiar pressure, but this pressure felt different. It wasn't crushing her; it was fueling her, giving her strength, reminding her of the years of training, the hours of dedication, and the unwavering support of her family. She understood this concerned Whisper also, not just herself. It was about their shared journey, their mutual trust.

At first, the jumps tested both skill and nerve. Leah concentrated on her technique and her bond with Whisper, pushing aside the self-doubt that threatened to take over. Whisper responded with precision and grace, her movements powerful and her leaps elegant. Each clear jump felt like a small victory, proof that she could succeed. The crowd's cheers lifted her, driving back the fear and doubt.

Then came the center fence—a formidable obstacle, its height and width a daunting challenge even for seasoned riders. It was the moment of truth, the test that would determine her fate. Leah's heart pounded in her chest, a frantic drumbeat against the silence of her concentration. She saw Emma watching her, and a determination coursed through her veins. It wasn't about Emma anymore. It was about proving herself.

As they approached the jump, Leah felt a surge of adrenaline, a potent cocktail of fear and exhilaration. She focused on Whisper, feeling the rhythm of her breath and the tension in her muscles. They were one, a single entity, poised for flight. She took a deep breath, trusting her instincts, training, and horse. She gave Whisper her cues: a subtle shift in her weight, a slight change of her reins. Whisper responded instantly, her powerful muscles propelling them forward. They soared through the air, a graceful arc against the backdrop of the cheering crowd, a testament to their skill and unwavering partnership. They landed perfectly; Whisper's powerful legs absorbed the impact.

The relief washed over Leah, a wave of pure joy. She had done it. She had faced her fears, overcome her doubts, and emerged victorious. With the weight of expectation gone, she felt a profound sense of accomplishment blossoming within her, a quiet confidence replacing the previous pressure. The center fence, a physical manifestation of her earlier anxieties, now stood tall and resolute, a powerful symbol of the courage and inner strength she had discovered within herself.

As she celebrated her victory, the crowd roared in approval, their cheers echoing and magnifying her triumph. She had cleared the jump with effortless grace, a display of her exceptional equestrian skill. When the competition ended, it felt almost dreamlike.

# Chapter 5: First Round Jitters

The initial jump was not a catastrophic fall, but a jarring jolt that sent a ripple of panic through Leah. Whisper, usually so reliable, hesitated slightly at the approach, a tiny stumble that, while imperceptible to the casual observer, felt like a monumental failure for Leah. The perfect rhythm, the effortless grace she'd strived for, had fractured, leaving her vulnerable. A tiny error triggered a flood of self-doubt.

Her heart hammered against her ribs, a frantic drumbeat that echoed the rising panic in her mind. The weight of expectation, previously a manageable burden, now felt crushing, threatening to suffocate her. The confident facade she had painstakingly constructed crumbled, leaving her exposed, raw, and acutely aware of the judging eyes fixed upon her. She could almost feel the weight of her family's legacy pressing down on her, the silent expectation of excellence hanging heavy in the air.

The next few jumps were a blur of frantic effort. She forced herself to focus, breathe, and reconnect with Whisper, but the earlier hesitation had broken the spell of calm concentration. Each jump felt like a monumental hurdle, each fence symbolizing her burgeoning fear. The smooth, practiced, jerky, hesitant actions, her body betraying the tremor of anxiety that racked her, replaced movements.

Sensing her rider's distress, Whisper responded with a subtle stiffness, her usual fluidity replaced by a hesitant gait. The once-perfect partnership, known for its seamless connection, appeared to be faltering, its delicate balance disrupted by Leah's inner turmoil. The usually comforting rhythmic thud of hooves striking the ground now relentlessly pounded in his ears, a constant, unwelcome reminder of the ticking clock, the ever-increasing pressure, and the genuine possibility of utter failure.

What had once been a source of encouragement, the crowd's murmurings, now felt like a chorus of judgment. This cacophony amplified her fears and highlighted every imperfection, making her increasingly self-conscious and exposed. With an almost painful intensity, she was acutely conscious of every single glance directed her way, every murmured word, and every minute alteration in the surrounding environment. Having once been a familiar and easily navigated challenge, the course now seemed alien and treacherous, a landscape filled with hidden dangers that were previously unknown. The showgrounds' exciting energy transformed; a heavy, stifling pressure pervades.

Leah clung to Whisper, seeking solace in the warmth of her horse's body, in the rhythmic beat of her heart against her leg. Whisper seemed to offer silent reassurance, a gentle nudge, a soft breath against her skin, a silent promise of unwavering support. It offered minor solace, preventing immediate collapse. Her mind raced, a whirlwind of negative thoughts and self-criticism. She repeated the initial mistake, dissecting her actions, searching for flaws, and magnifying imperfections. The voices of doubt, so quiet only moments before, had now grown into a deafening chorus, whispering insidious lies that undermined her confidence and chipped away at her resolve.

However, within the confusion, a rebellious spirit emerged. Despite the overwhelming fear, she resolutely refused to let it negatively affect her performance. Despite the immense pressure she was under, she resolutely refused to yield. She recalled her training, her family's unwavering support, years of dedication, and the ride's pure joy. These memories empowered her, as she concentrated on riding and her bond with Whisper.

Relief mingled with anxiety as Leah faced the next few fences. She grounded herself in the basics, focusing on technique and steady breathing. With each stride, she sought to restore the rhythm, harmony, and unspoken connection that defined their partnership. Every clear jump became a small victory, a step toward rebuilding her confidence and proof that she could still perform under pressure.

The course progressed; each jump a new battle against her self-doubt. There were moments of brilliance and effortless grace, but also moments of hesitation and near misses, each a testament to the emotional turmoil raging within her. The weight of expectation continued to press down on her. Still, with each successful jump, the burden felt slightly lighter, the pressure marginally less intense as she approached the last hurdles. An unexpected tranquility washed over Leah. Having confronted her fears and acknowledged the doubts that plagued her, she emerged from the ordeal, bearing the marks of her struggles, but ultimately unbroken in spirit. No longer bound by the relentless pursuit of flawlessness, she freed herself from its constraints.

As the competition reached its climax, the athletes braced themselves for the last jump. The aim wasn't to achieve a perfect performance; it was about something else entirely. This experience involved confronting the pressure, overcoming the jitters and internal chaos, and maintaining control and composure. This endeavor served as a personal challenge, a testament to her determination to conquer adversity. With a newfound clarity, Leah guided Whisper forward, her actions imbued with confidence and purpose. Moving as one, a restored rhythm, a shared goal giving them both focus and strength, marked their synchronized movements. With a resounding thwack, they cleared the jump, and the echo of the sound mirrored the wave of released tension that washed over her. The course was complete.

Exhaustion mixed with relief washed over her as she brought Whisper to a halt, the emotion of the competition still pulsing through her veins. It wasn't perfectly round, and it was far from it. Imperfections were clear because of mistakes she had made. But she had completed the course. She had faced her fears. She had persevered. The result wouldn't matter as much as her resilience and bravery. She had faced the weight of expectation and emerged, not necessarily unscathed, but stronger for it. The first-round jitters had been intense, almost unbearable. However, this experience strengthened Leah, teaching her true courage. The journey had begun, and Leah passed the first test. She rode away from the first round, knowing she would be prepared next time.

# Chapter 6: Analyzing the Course

The course map spread before Leah, a stark white sheet contrasting with the vibrant colors of the arena beyond her window. She traced the lines with a fingertip, her brow furrowed in concentration. The map meticulously detailed each jump, a miniature representation of the obstacles ahead. This wasn't a visual representation; it was a strategic puzzle, a challenge to be dissected and conquered.

Her gaze lingered on the 'center fence,' a triple-bar jump marked with a bold, menacing red. It was the crux of the course, a formidable obstacle that tested not technical skill but also nerve and mental fortitude. The height wasn't excessively daunting, but the triple bars demanded precision, a perfect distance, and an unwavering rhythm. One misplaced stride or hesitant moment, and the entire sequence could unravel, leading to a costly fall.

She remembered the whispers from other riders—"The Center Fence Breaker," some called it, an innocent phrase that held a chilling undertone. She'd seen seasoned competitors struggle with it, their carefully planned approaches undone by sudden hesitation, a loss of confidence, or a lapse in connection with their horse. One wrong jump ruined a flawless round, undermining even seasoned riders' confidence.

Leah, however, saw not a threat, but a challenge. Despite the setback, she refused to let it defeat her spirit, and she resolved to overcome it. With unwavering resolve, she knew she had the strength and determination to rise above and master it. In her analysis of the approach, she meticulously considered the distance required for the jump, paying close attention to the angle of the preceding turn, the quality of the footing, and the subtle incline of the ground. Every single detail was of critical importance, and each contributing factor played a significant role in the development and execution of the overall strategic plan. In her mind, she meticulously played back the sequence, creating a vivid image of every stride Whisper would take, and carefully predicting every movement in advance.

Preparing for the challenge ahead required more than physical fitness; mental fortitude was just as essential. Yet despite her efforts, her thoughts wandered, drifting far from her immediate surroundings. As she pictured the scene in her mind, she could almost feel the powerful muscles of Whisper beneath her, sense the steady rhythm of their combined strides as they moved in perfect unison.

She vividly recalled the failures of the last round. Confidence had been fragile, and self-doubt had held too much power. Yet the memory was not a burden but a lesson and a catalyst for growth. She refused to let past mistakes dictate her future performance. This time, she approached the challenge differently, armed with a newfound awareness and a steely determination to conquer her fears.

She picked up her pencil and made small annotations on the map—the optimal stride length, the correct turn point, and the ideal approach speed. Her notes reflected a meticulous analysis, a blend of theory and experience, a testament to her dedication and passion for the sport. Every stroke of her pencil showed her. She is committed to overcoming the obstacles ahead—both the physical challenges on the course and the mental ones within herself.

Meticulous project planning prioritized a complex, expansive design, ignoring the center fence. Analyzing each jump individually, she meticulously examined the distances covered, the angles of approach and departure, the obstacles encountered, and the smooth transitions or potentially problematic shifts between them. In her mind's eye, she pictured the graceful and fluid progression of the course, the seamless and harmonious rhythm of Whisper's strides, and the flawless unity and coordination between herself and her equine partner.

Before each jump, she meticulously developed and reviewed a concise mental checklist that included the following phases: her approach to the jump, her stride, the takeoff, the flight itself, and finally, the landing. She didn't passively review the course; instead, she engaged with the material, asking questions and seeking clarification where needed. In her assessment of the course design, she pondered the nuances of its subtleties, including the strategic positioning of the jumps, as well as the difficulties presented by the turns and transitions between different sections of the course. The course designer, a person of undeniable skill, had clearly crafted a test that would challenge both the technical abilities and the mental resilience of the participants. The jumps were physically demanding and strategically placed to challenge a rider's focus, rhythm, and timing.

Leah felt a surge of excitement mingled with a healthy dose of apprehension. This wasn't another competition; it was a test of her resilience, a crucible where she could forge a stronger, more confident version of herself. The fear was still there, a familiar shadow lurking at the edge of her awareness, but it no longer held the same paralyzing power.

It had become a sparring partner, a worthy opponent to be challenged and ultimately defeated. This wasn't about avoiding fear but embracing it, learning to navigate it, and ultimately using it as fuel to propel her forward. She studied the course again, this time with renewed confidence. The center fence remained a significant challenge, but it was no longer a daunting obstacle; it was a goal, a test to be overcome. Her strategy was simple, her focus sharp, and her determination unwavering.

She looked out the window again, her gaze fixed on the brightly colored jumps gleaming under the stadium lights. The arena's energy was palpable, a vibrant buzz that excited and invigorated her. She pictured herself and Whisper moving through the course, a harmonious team, each movement fluid and precise. Because of the rhythm of their partnership and their unspoken communication, which served as a silent promise of success, they would overcome the challenges that lay ahead.

Despite immense pressure, she'd learned to transform it into motivation. Driven by family, coach, and self-imposed expectations, she persevered. They reminded her of her dedication, years of hard work, and self-belief, highlighting her achievements. After reviewing the course map, Leah carefully folded and safely stored it in her backpack to protect it. She had mentally prepared herself. Leah had carefully performed the tactical analysis.

The only thing left was for her to carry out her plan; she would need precision, courage, and unwavering focus. Leah, confident and determined, prepared to meet the challenge. Next, she concentrated on physical riding, mental strength, and overcoming inner demons—a struggle to reflect significant upcoming challenges. Her true skill wasn't jumping fences, but overcoming self-doubt and fear. Leah knew she'd discover her true strength within that experience.

# Chapter 7: The Pressure Mounts

The low hum, a background thrum all day, intensified as the last competitors finished the course. The crowd applauded each cleared fence, creating a wave of sound that crashed against the arena's walls and echoed in Leah's ears. But the silence that followed a fall, the hushed expectancy, was far more potent. It hung heavy in the air, a palpable tension that tightened the knot in Leah's stomach.

She watched, her heart pounding a frantic rhythm against her ribs, as seasoned riders, known for their skill and composure, faltered at the center fence. One rider, a competitor she'd admired for her effortless grace, clipped a rail, the jarring sound sending a shiver down Leah's spine. Another, a powerful rider with a reputation for taking risks, came too close, narrowly avoiding disaster but losing valuable seconds. The center fence, that seemingly innocuous-looking triple bar, was proving to be a test of mettle, a crucible refining the courage of even the most experienced competitors.

The spectators' whispered comments reached her—snippets of conversation that pierced the concentration she was attempting to maintain. She heard someone said, "One more down at the center." Another murmured, "That fence is a nightmare." Each fallen rail, each hesitant approach, fueled the growing anxiety within her. The pressure wasn't the physical demands of the jump; it was the weight of expectation, the silent judgment of the crowd, and the knowledge that one mistake could unravel everything for which she had worked.

The stakes were higher than ever before. This wasn't a local competition but a qualifier for the regional championships—a stepping stone towards her goal. Years of training, countless hours spent honing her skills, countless sacrifices—it all culminated in this moment, in this single round, on this unforgiving course. A perfect run mattered, not scores, but self-affirmation.

The thought of failure, of falling short of her expectations and those of her family, weighed heavily upon her.

She ran her fingers across Whisper's soft muzzle, feeling her heart's warm, rhythmic beat. The mare remained calm, reassuring Leah amid the emotions raging within her. Whisper sensed her rider's anxiety and slightly tense posture but offered no judgment, only unwavering support. Their silent communication was a comfort, a reminder of their enduring bond.

The anticipation was almost unbearable. The announcer called her name, cutting through the crowd's murmur. Leah's breath caught, and for a moment she froze. Then she remembered her training, the hours of mental preparation and countless visualization exercises. She inhaled deeply, focusing on the steady rise and fall of her chest, anchoring herself in the present.

She mounted Whisper, the familiar feel of the saddle a grounding comfort. The mare shifted under her, sensing her rider's readiness. Leah took a moment, a few seconds, to connect with Whisper, reinforce their shared understanding, and feel the powerful, steady energy emanating from the magnificent creature beneath her. It was a moment of silent communion, a pact between horse and rider, a promise of mutual support and trust.

Entering the arena, light gleamed off her polished saddle; the moment's weight intensified. As she moved forward, the judges intently watched her every move. The crowd held its breath in anticipation, and the ominous shadow of the center fence loomed large before her. Although the fear remained, its nature had shifted; it no longer induced paralysis, but manifested as a concentrated, almost acutely defined energy. The course unfolded before them, with each jump a challenge and each turn a strategic decision. Leah rode with precision and grace, guiding Whisper through the sequence with an almost supernatural fluidity. They moved as one, a synchronized team, perfectly executing each stride and jump. The rhythm was perfect. Every movement was effortless.

The energy felt almost mystical, a silent conversation between horse and rider. They were a unit, a single entity working towards a singular aim. She easily cleared the first few fences, her confidence growing with each successful jump. With each jump, she felt a rising sense of exhilaration. Whisper responded to her every touch and subtle cue with effortless grace and power. They were a team, perfectly in sync, a harmonious blend of strength, skill, and intuitive understanding. And even though the pressure remained immense, Leah moved with a newfound calm.

As she approached the last jump, the center fence, a wave of anxiety washed over her once more. But this time, the anxiety wasn't debilitating; it was fuel, a potent stimulant pushing her forward. She focused on the approach, her eyes fixed on the triple bar, visualizing the perfect stride, the precise takeoff, and the flawless landing. The distance was perfect, the rhythm impeccable. Whisper responded to her commands without hesitation, her powerful strides carrying them towards the fence with unwavering confidence. With Whisper's thrust, Leah felt the lift; time stood still. They soared over the obstacle, landing smoothly and effortlessly. A collective sigh of relief washed over the arena. Leah had conquered the center fence.

The remaining fences were almost an afterthought. Her victory over the center fence shattered her doubts and boosted her confidence for the rest of the course. She rode with renewed determination, each jump a testament to her growing self-belief. She crossed the finish line, victorious. The applause that followed was deafening, a wave of sound that washed over her, a testament to her triumph. She had cleared the course and conquered her fears, proving that she could achieve anything she set her mind to. The exhilaration was immense, a triumphant feeling that surged through her veins. The pressure had been enormous, the stakes high, but she had risen to the challenge, transforming anxiety into strength and fear into determination.

This victory was not merely a competitive win; it was a triumph over personal challenges, a powerful testament to her unwavering perseverance, and a moving demonstration of the deep connection she shared with her horse. Although the pressure may have intensified, she persevered, rising above the challenges to emerge stronger, more self-assured, and better prepared than ever before to tackle any future obstacles that might arise. It was an intoxicating feeling, a powerful culmination of years of dedication, tireless hard work, overcoming considerable self-doubt, and ultimately rising to meet the challenge. She had arrived at a pivotal moment, one of profound significance that would forever remain etched in her memory as a treasured experience.

# Chapter 8: A Moment of Doubt

The crowd's roar faded into a dull hum, the individual cheers blurring into an indistinguishable drone. Leah sat on Whisper, her heart hammering a frantic rhythm against her ribs, the triumphant feeling of clearing the earlier fences dissolving into a chilling wave of self-doubt. The exhilaration had been fleeting, replaced by a bone-deep fear that threatened to paralyze her. The center fence, conquered moments ago, now loomed in her memory, not as a testament to her skill, but as a terrifying reminder of how easily it could all come crashing down.

It wasn't the physical challenge; it was the mental strain, the sheer weight of expectation that pressed down on her, threatening to suffocate her. She thought of her family, their unwavering belief in her, and the legacy they expected her to uphold. The pressure was immense, a crushing weight that threatened to break her. The whispers of doubt, once barely audible, now echoed loudly in her mind, amplifying her fears. What would happen if, for whatever reason, she could not sustain her current level of high performance? Imagine if, in a moment of weakness or hesitation, she were to stumble or lose her footing, maybe even fail completely. What if, by some unfortunate turn of events, her actions resulted in disappointing every one of them?

The thought of failure gnawed at her, devouring her confidence. The memory of the previous jumps, their effortless grace and flawless rhythm, seemed impossibly far away, like a mirage in the harsh desert of her current fear. She felt disconnected, adrift in a sea of self-doubt. The comforting presence of Whisper suddenly felt distant, the bond between them strained by the intensity of her inner turmoil.

A wave of nausea washed over her. Her hands, usually steady and controlled, trembled slightly on the reins. She could feel the tension radiating through her body, tightening her muscles, and making her stiff and inflexible.

She pictured the next fence; it wasn't a challenge, but an insurmountable obstacle. As I looked, the highly polished wood, and the perfectly placed rails seemed to blur, taking on a dark and foreboding appearance, becoming ominous and unclear.

Quitting the competition became increasingly tempting. That option would represent the simplest solution, requiring minimal effort and presenting the path of least resistance. By avoiding failure, she could escape the humiliation it would bring, prevent her family's disappointment, and ease the crushing weight of expectation that bore down on her. Although the thought took hold in her mind, a powerful resistance ignited within her, a strong and unwavering refusal to succumb to the grip of fear.

She closed her eyes, trying to regain her composure and summon the strength she knew she possessed. She breathed deeply, trying to calm the erratic beating of her heart. The image of her father, his unwavering support and patient guidance, flashed before her eyes. She remembered his words, encouragement, and faith in her abilities. They were a lifeline, pulling her back from the brink of despair. But the doubts lingered. The memory of past failures, those moments of weakness that had plagued her throughout her career, surfaced. They were ghosts haunting her, whispering insidious lies in her ear. They reminded her of her vulnerability, imperfections, and capacity for error. Each doubt chipped away at her confidence, slowly, relentlessly.

She opened her eyes, looking at Whisper. The mare stood patiently, her head lowered slightly, sensing her rider's distress. Their eyes met, and in that moment, Leah found a sliver of solace and hope. Whisper's unwavering calm and silent trust was a balm to her troubled soul. It evoked memories of their past, highlighting the strong bond formed through years of training and competition. The competition wasn't the point; Abandoning the project meant betraying their mutual trust. Responsibility for her horse reignited her determination, filling her with renewed purpose.

Countless hours of training, marked by sweat, blisters, and unwavering dedication, came rushing back, bolstering her resolve and fueling her commitment. Despite the nagging voice of self-doubt, she refused to let it eclipse the dedication, relentless training, and unwavering commitment she had poured into her preparation for so many years.

The announcer's voice pierced through the silence of her internal battle. It was a jarring sound, pulling her out of her contemplation and grounding her in the present moment. Leah took a deep breath, focusing on her chest's rhythmic rise and fall and the steady beat of Whisper's heart against her leg. The fear was still there, a persistent hum beneath the surface, but it no longer held the power to paralyze her.

She knew the next few jumps would be a test of her mettle. It would focus on resilience, overcoming discomfort, and her ability to handle internal and external challenges. The path ahead was difficult, but Leah felt an unexpected surge of resolve; she was ready to face whatever came next. The fear remained, but it no longer controlled her. It was merely a companion, a shadow that followed her, but one she could now live with and perhaps harness. The competition was not against other riders; it was, ultimately, a battle against herself and her demons. And in that moment, she was ready for the fight. The rhythmic thud of Whisper's hooves broke the silence of the arena only as Leah turned, preparing for the next challenge. Despite a fence victory, an actual courage test awaited. The journey of self-discovery had only begun.

# Chapter 9: Words of Encouragement

The tension in the air was palpable, thick enough to cut with a knife. Murmurs and nervous energy from rivals created a heavy atmosphere. Leah felt a familiar wave of nausea rise in her throat, the fear threatening to overwhelm her once more. She adjusted her helmet, the smooth leather a cold comfort against her clammy skin. Whisper, sensing her rider's distress, shifted restlessly beneath her.

Then a hand rested lightly on her shoulder. Leah jumped, startled, and turned to see Emma, her usually aloof and competitive rival, standing beside her. The unexpected touch and presence broke through the wall of fear that had been slowly closing in.

Surprisingly, Emma offered a small, hesitant smile. "You were incredible over the center fence," she said, her voice surprisingly soft and devoid of the usual arrogance Leah expected. I saw it. You nailed it."

Leah stared at her, speechless. Emma's words were a balm to her wounded spirit, an unexpected lifeline in the turbulent sea of self-doubt. She had expected criticism, perhaps even a condescending remark about her earlier hesitation, but this... this was different.

"Thanks," Leah whispered, her voice barely audible. The sincerity in Emma's eyes surprised her. It wasn't the calculated politeness of a competitor feigning civility, but a genuine expression of admiration.

Emma shifted her weight, her usual confident posture slightly subdued. "It's difficult, you know," she said, her voice dropping to a near murmur. "The pressure... It's a beast. It can eat you alive if you let it."

Leah nodded, surprised by the vulnerability in Emma's words. It starkly contrasted the image she had carefully constructed of her rival—a seemingly invincible force, unburdened by doubt or fear. The revelation chipped away at the carefully constructed walls of Emma's persona, revealing a fellow competitor who struggled with the same demons.

"I've been there," Emma continued, her eyes drifting to the course ahead. "I've had rounds where I've completely choked, where the fear took over. It's humiliating, terrifying. You feel you're drowning, and there's nothing you can do to save yourself."

Leah listened intently, finding a strange comfort in Emma's honesty. It was a shared experience, a secret bond forged in the crucible of intense competition and crushing self-doubt. The weight of her fear suddenly felt a little lighter, a little less isolating. Emma leaned closer, lowering her voice even further. "But you know what?" she said, a hint of defiance creeping into her tone. "Fight back. Fight the fear. You can't let it win."

Her words resonated deeply with Leah. They were not merely words of encouragement, but a battle cry, a call to arms. Emma's acknowledgement of her struggles and honest admission of vulnerability created a space for Leah to confront her fears without shame.

"How?" Leah asked, her voice trembling slightly.

Emma smiled, a genuine, warm smile that reached her eyes. "Focus on your horse," she said. "Feel the rhythm, the connection. Trust your instincts. And remember why you're here. Remember the hours you've spent training, the dedication, the passion."

Emma's words were simple, yet profoundly influential. They reminded Leah of the fundamental principles of riding, the importance of the partnership between horse and rider, and the unwavering trust and connection that formed the bedrock of their success.

"And remember," Emma added, her voice dropping again to a whisper, "it's okay to be scared. Fear is normal. It's how you deal with the fear that matters."

Leah felt a surge of gratitude, a wave of warmth washing over her as she absorbed Emma's unexpected kindness. The weight of expectation, the crushing pressure, seemed to ease, replaced by a newfound sense of camaraderie, a shared understanding of the intense emotional toll of competitive riding.

Emma's words were more than encouragement; they validated her feelings, allowed her to be vulnerable, and acknowledged her fears without shame or self-recrimination. The unexpected support and friendship from the least likely source helped change everything. With this new perspective, once seemingly insurmountable, the competition now appeared less daunting, its obstacles no longer insurmountable. Although the fear had not wholly disappeared, its monstrous and paralyzing presence was no longer as overwhelming as it once was.

Spending time with Emma reminded me that even in the harsh world of competitive horse riding, it's essential to be kind and connect with others. Because of their shared vulnerability and the unspoken understanding that grew between them, Leah's resolve strengthened, and she found a renewed confidence that extended far beyond the immediate comfort provided by this surprising ally. With a newfound determination, born from her resilience and the surprising competitor's support, she eagerly expected the challenges that lay before her.

The close bond she'd carefully built with Emma, even with their rivalry, provided immense emotional support, helping her face the competition's challenges. Despite lingering fear, a shadowy presence at the edge of her awareness, it no longer held power over her. Whisper's renewed bond with the narrator, strengthened by her restored confidence, empowered her to overcome the last obstacles with revitalized energy. As she jumped repeatedly, the challenge seemed less daunting, the task's intimidating nature fading as the arena's atmosphere noticeably changed, lightening her burden. The unexpected friendship and small camaraderie that had formed had given Leah the extra support and encouragement she required to succeed.

The following jumps were not perfect, but they were strong. Each one stood as a testament to her growing resilience. Fear lingered, a persistent undercurrent, but it no longer controlled her. She focused on Whisper, on their shared rhythm and the trust built over years of partnership. Pushing through discomfort and anxiety, she drew strength from Emma's words and the unexpected support that had pulled her back from the edge of despair.

Each fence cleared was a victory, not against the course, but against the doubts that had plagued her. Each jump was a step towards self-acceptance and a more honest, compassionate view of her capabilities. Leah's grace and self-assurance were surprising as she easily conquered the immense, previously daunting jump.

The crowd's cheers felt different as she rode towards the finish line — less intimidating and more celebratory. Leah could feel the weight of her accomplishment lifting, replacing the oppressive fear with a sense of achievement, growth, and self-discovery. The journey hadn't been easy, but she had persevered, overcoming not the challenges of the course but the even greater challenge of conquering her inner demons. Emma's quiet words of support, offered amidst the chaos of the competition, had become the catalyst for her transformation, a turning point in her personal and athletic journey.

The unspoken camaraderie between competitors, the mutual understanding of the pressures and anxieties inherent in their sport, had proven to be a powerful force. The fear was not gone entirely, but it was now a manageable companion, a shadow that accompanied her but could no longer enslave her. She had found her strength, not in the absence of fear, but in her ability to embrace, understand, and ultimately overcome it. And that, perhaps, was the most significant victory of all.

# Chapter 10: Embracing the Challenge

The air crackled with anticipation. The rhythmic thud of hooves on the ground, the hushed whispers of the crowd, the scent of freshly turned earth and horse sweat—it all swirled around Leah, a vortex of sensory overload that threatened to pull her under. But this time, the fear, while present, felt different. It was an icy knot in her stomach, a tightening in her chest, but it didn't hold the exact paralyzing grip it had before. Emma's words, her unexpected vulnerability, had created a space, a crack in the wall of her self-doubt.

As Leah's demeanor shifted, Whisper, sensing the change in her rider's mood, stood patiently, her soft breath a gentle caress against Leah's leg. Leah felt a quiet strength, and the mare seemed to sense it. It wasn't a lack of fear, but a newfound understanding. The challenge wasn't an enemy to defeat, but something to overcome.

Looking ahead, the center fence appeared a formidable obstacle, its height and width emphasizing its intimidating presence. This wasn't merely a test of physical prowess and a trial of mental resilience, demanding that participants transcend their fears and insecurities. Leah's past mistakes echoed faintly, fueling her self-doubt. However, the voices were quieter today, subdued by a newly discovered determination.

She took a deep breath, focusing on the rhythm of Whisper's breathing and the steady beat of her heart against her leg. She felt the familiar connection—the unspoken language between them, the mutual trust forged through years of training, shared victories, and defeats. This wasn't about clearing a jump, but reaffirming their partnership and proving that she could trust Whisper and her capabilities.

Leah guided Whisper towards the approach, her body relaxed yet alert, her movements fluid and precise. With meticulous care, she measured the length of each stride, ensuring that each movement was deliberate and accurate. Rather than concentrating on the potentially disastrous height of the fence, she focused on the plan's details and its careful execution. As they moved together, she intently observed Whisper's gait, meticulously maintaining their shared rhythm while simultaneously interpreting the subtle shifts in the mare's body language that revealed their unspoken connection and understanding.

As they approached, a surge of adrenaline coursed through them, yet, unlike before, it wasn't the familiar, crippling wave of panic. Leah could feel the potent energy, the raw strength and athleticism, emanating from Whisper, a power coupled with the animal's absolute, unquestioning trust in her direction and expertise. Their collaboration, a kind of dance, beautifully showcased the years of commitment and rigorous training, culminating in this pivotal moment.

Then, at the fence's base, loomed the imposing obstacle. Leah took another deep breath, the air filling her lungs and renewing her focus. She didn't think about falling or dwelling on the possibility of failure. She thought about the connection with Whisper, the unwavering trust they shared, the years of sweat and toil, and the shared victories and crushing defeats that had bound them together.

Whisper responded instantly to the slightest shift in her weight and the almost imperceptible adjustments of her posture. It was a seamless exchange, a perfect harmony of body and mind, the epitome of the horse-rider partnership. Leah softly urged Whisper forward, her voice barely audible above the mare's hoofbeats.

The takeoff was clean; the whisper soaring effortlessly, her muscles rippling, her movements fluid and precise. Leah felt the familiar sensation of flight, the world falling away as they arced through the air, time seeming to slow as they hung suspended between earth and sky. Then, the landing, gentle and precise, and then the exhilaration of another fence cleared.

The relief that washed over her went beyond finishing the jump; it was a deeper, more profound sense of accomplishment.

The rest of the course was a blur, a sequence of precise movements, fluid transitions, and unwavering focus. Each cleared jump was a testament to her newfound resilience, confidence, and ability to push beyond her limitations.

As they crossed the finish line, the crowd's cheering was deafening, a wave of sound that washed over her, a testament to her accomplishment. Leah sat tall in the saddle, her chest swelling with pride, not for her performance but for her ability to overcome her fear. She had not only cleared the center fence but had transcended it, transforming a symbol of her anxiety into a monument to her triumph.

The journey wasn't easy, and the challenges were not simple, but Leah persevered. She faced her inner demons and emerged victorious. The flawless execution of her course did not define this victory, but by the quiet, internal battles she fought and won within herself. She embraced the challenge, overcame her self-doubt, and proved to be stronger, braver, and more capable than she had ever imagined.

I felt a profound sense of accomplishment, a deep and resonating satisfaction that transcended the competitive environment and filled me with a joy that extended far beyond the competition's end. More than just winning, this success resonated within her. It was far more than a simple show jumping competition; this event encompassed a wider range of activities and elements, transcending the boundaries of a typical equestrian competition.

The subsequent rounds were solid, efficient, and confident. The fear lingered, a whisper at the edge of her awareness, but it no longer held power over her. It was a reminder, a shadow, but it no longer dictated her performance. She maintained her composure, focusing on the precision of her movements and the harmony between her and Whisper. Leah performed each jump with unexpected precision and control.

The competition's last jump, a daunting obstacle, initially appeared unconquerable. Now, facing it, Leah felt a rush of calmness rather than panic. She had overcome her biggest internal challenges, proving her ability to manage fear. She approached it with a newfound serenity, drawing on the strength she'd discovered within herself. Whisper responded, their partnership honed to perfection by the shared experience of overcoming adversity. They cleared the final fence quietly, symbolizing their collective victory.

The cheers seemed less intimidating and more celebratory as they exited the arena. The weight of expectations had lifted, replaced by a quiet satisfaction and a profound sense of accomplishment. Leah had faced the challenge, embraced the fear, and emerged triumphant. The journey, a transformative odyssey fueled by courage, resilience, and an unwavering spirit, had been one of profound self-discovery.

# Chapter 11: The Approach

The sun beat down on the arena, the dust motes dancing in the golden light. The crowd, a sea of expectant faces, hummed with a low, anticipatory thrum. But Leah barely registered them. Her world had shrunk to the precise dimensions of the arena, to the rhythmic beat of Whisper's hooves, to the steady rise and fall of her breath. The center fence, the infamous obstacle that had haunted her dreams and stolen the sleep from her eyes for weeks, loomed ahead. It wasn't a jump, but a symbol — a monument to her past failures, a stark reminder of the self-doubt that had clung to her like a shadow.

Whisper, her magnificent chestnut mare, shifted beneath her, her weight settling into a comfortable, ready stance. The mare's warm breath against Leah's leg was a soothing counterpoint to the frantic beat of her heart. Years of training, of shared victories and heartbreaking defeats, had forged a bond between them, an unspoken language of trust and understanding. Equal partners, Leah, felt unwavering faith in their shared skill and courage.

She adjusted her grip on the reins, feeling the silken texture against her gloved hands. She didn't focus on the height of the fence, the treacherous width, or the potential for disaster. Instead, she focused on Whisper's breath, the steady rhythm of her strides, and the subtle shifts in her body language. It was a delicate dance, a conversation conducted in the unspoken language of the horse and rider, a symphony of muscle memory and instinctive understanding.

The approach was crucial. Too fast, and they'd overshoot, losing momentum and risking a disastrous fall. Too slow, Whisper would lose her rhythm, spring, and chance of clearing the fence cleanly. Leah guided Whisper with barely perceptible adjustments of her weight, her body a conduit for their unspoken communication. They measured each stride and made each movement precise; their choreography, honed by countless hours of practice, showed their unwavering connection.

The scent of freshly turned earth and horse sweat filled the air, a heady perfume that spoke of exertion and anticipation. The crowd's whispers faded into the background, replaced by the rhythmic thump of Whisper's hooves and the steady beat of their combined heartbeats. Leah closed her eyes briefly, silencing the internal critics and pushing aside the ghosts of past failures. She breathed deeply, focusing on the rhythm, the feel of Whisper's muscles beneath her, and the power and grace between them.

When she opened her eyes, the center fence was almost upon them, its imposing presence filling her vision. The air crackled with anticipation, and a palpable energy filled the arena. But Leah didn't flinch. She had faced her fear, confronted her self-doubt, and emerged stronger and more resolute. This wasn't about clearing a fence but proving to herself that she could conquer her inner demons and rise above her limitations.

This time, adrenaline-fueled excitement, not fear. It propelled them forward, fueling their movement and amplifying their connection. She felt Whisper's power, the mare's willingness, and her unwavering trust. There was no hesitation, no second-guessing, only the instinctive precision of years of training, the seamless partnership born of mutual respect and tireless dedication.

Leah urged Whisper forward, her voice a gentle murmur, barely audible above the rhythmic thud of hooves. The mare responded instantly, her body a fluid extension of Leah's will, perfectly reflecting her rider's intent. They approached the fence with a steady gait.

The takeoff was perfect. Whisper launched into the air, her muscles rippling with power, her movements fluid and precise. Time seemed to slow as they arced gracefully through the air, briefly suspended between earth and sky. The world fell away, leaving only the intimate connection between horse and rider, the shared breath, and the mutual trust.

The landing was equally flawless, gentle, and controlled. Whisper landed with barely a tremor, her weight evenly distributed, her balance unyielding. For a breathless moment, they stood poised, still; the silence broken only by the muffled sounds of the crowd. Then, Leah spurred Whisper forward, and they continued their course, the rest of the jumps a blur of focused movements and effortless grace.

The relief that washed over Leah wasn't simply the relief of a completed jump, but a deeper, more profound sense of accomplishment. Her impressive athleticism and skill, clear in clearing the center fence, left the crowd speechless. Having confronted her most crippling doubts and having faced her deepest fears, she had emerged from the ordeal victorious. The remaining jumps seemed more straightforward, less daunting. The fear was still there, a faint undercurrent, but it no longer controlled her. It was a shadow, a whisper, but it no longer dictated her actions. Each jump became a testament to her newfound resilience, burgeoning confidence, and ability to push beyond the boundaries of her self-doubt.

As they crossed the finish line, the crowd's roar was deafening, a wave of sound that washed over her, celebrating her achievement. But Leah heard only the quiet rhythm of Whisper's breathing, the steady beat of her heart, the gentle whisper of her triumph. She had proven to herself, more than anything else, that she was capable of greatness, could overcome her fears, and could trust both her horse and herself.

The victory resonated far beyond the confines of the competition arena, extending into the deepest recesses of her being. This was more than a show jumping competition; it was a journey of self-discovery, a testament to the resilience of the human spirit, a profound affirmation of her strength and courage. The ribbons and rankings paled compared to the quiet satisfaction she felt deep within, the unshakeable conviction that she could face any challenge, overcome any obstacle, as long as she remained true to herself, to her partnership with Whisper, and her unwavering belief in her capabilities. The journey was far from over; this was only the beginning. She had cleared the center fence and so much more with it.

The path ahead remained challenging, but now she knew she had the strength, resilience, and self-belief to face it head-on. And she would continue to ride, jump, and strive for excellence, drawing strength from the victory she had already won. The memory of that moment, the sensation of soaring over the fence, and the unwavering trust with Whisper would always be her touchstone, offering strength and confidence for upcoming challenges. One jump, one win, yet this win promised many more. This was a turning point, a new beginning.

# Chapter 12: The Jump

The world narrowed. It wasn't the roar of the crowd, the expectant hush before the storm, or even the imposing height of the center fence itself that filled Leah's consciousness. It was the precise rhythm of Whisper's strides, the subtle shift in her weight, and the almost imperceptible tremor in her muscles. This progressed not via a jump but through dialogue—a refined interplay of trust and understanding developed through years of mutual experience.

The approach was a blur of controlled motion. The rider meticulously measured each stride, barely perceptibly adjusting the reins. Leah felt Whisper beneath her, a powerful, willing partner. Her breath was warm against Leah's leg, a silent, reassuring touch. Spectators and competitors held their breath, anticipating the moment.

They were a unit, a single entity moving with breathtaking precision. Leah's body moved as one with Whisper, anticipating the mare's every shift and guiding her with barely a whisper. Intense concentration cut through the surrounding chaos, directly addressing the task—only the present, the impending leap, no room for doubt or fear.

Gradually, the distance that separated them became shorter and shorter. She now reached what had once been a massive obstacle, a looming symbol of her deepest anxieties and fears, no longer a source of dread but a challenge she would bravely meet and ultimately conquer—a test of her strength and resilience. The fence appeared to increase in size, its height and width growing astonishingly, expanding into something almost unreal in its vastness. Leah, however, stayed completely focused, her attention remaining as precise and unwavering as a laser beam. Because of the poor visibility, she could not see the fence itself.

Then came the takeoff. It wasn't a sudden burst of speed, but a controlled explosion of power and grace. Whisper surged forward, her muscles coiling and uncoiling with breathtaking power, her movements as fluid and precise as a dancer's. Leah felt the powerful thrust of her mare's hindquarters, the effortless lift, the almost weightless feeling of ascent.

They hung suspended in midair for a heartbeat, a breathtaking tableau of athleticism and grace against the vibrant arena backdrop. The world below was a dizzying swirl of colors and motion, yet Leah's focus remained utterly undisturbed. Only Whisper, wind, and breathing remained.

Leah felt an exhilarating rush of adrenaline, mirroring Whisper's powerful movements. It was a feeling of flight, freedom, and an unparalleled connection between horse and rider—a harmonious symphony of shared purpose and unwavering trust.

The landing was as smooth and controlled as the takeoff, a descent that barely disturbed the rhythm of their movement. There was no jarring impact, no stumbling, no loss of balance. Whisper landed with a barely audible sigh, her weight evenly distributed, her muscles relaxed and ready for the next challenge. The ground steadied beneath them, a firm anchor in the aftershock of their aerial ballet.

The silence that followed was profound, a brief pause before the eruption of cheers from the assembled crowd. Leah drew a deep breath, savoring the moment, the quiet triumph of a perfectly executed jump. She felt a wave of pure exhilaration wash over her, the tangible reward of countless hours of training, dedication, and unwavering perseverance.

It was more than a jump; it was a turning point. Once a daunting obstacle, the center fence now stood as a testament to her strength, courage, and unwavering bond with Whisper. Her longstanding nightmare faded; confidence and inner strength replaced it.

The remaining course flew by in a blur of motion. Previously daunting obstacles, the remaining fences appeared minor compared to the central challenge. Leah felt an unshakeable calm, a steady hand on the reins, and an unwavering trust in her mighty mare. The rhythm remained consistent, the focus unwavering, the movements precise.

As they crossed the finish line, the crowd's roar was deafening, a wave of sound enveloping them, celebrating their success. But Leah only heard the steady rhythm of Whisper's breath, a gentle counterpoint to the thundering applause. It was a victory, not only in the competition, but within herself.

The ribbons and accolades that followed felt almost secondary, a pleasant aftertaste to the potent brew of adrenaline and triumph that coursed through her veins. The genuine victory was the silent conversation between her and Whisper, the silent acknowledgment of their shared achievement, the unspoken pact of mutual trust and respect forged in the crucible of competition.

Later, as the initial euphoria faded, and the dust settled, Leah reflected on the significance of that single jump and its profound impact on her confidence and self-perception. It wasn't about clearing a fence but overcoming a deep-seated fear, confronting her self-doubt, and emerging victorious. The center fence had become more than a jump; it had transformed into a symbol of her personal growth.

The feeling of flight, the exquisite grace of the leap, and the unwavering trust between herself and Whisper forever remain etched in her memory. It was a poignant reminder of her capabilities, resilience, and the profound bond between horse and rider—a bond forged in training and shared experiences of triumph and adversity.

The journey ahead would undoubtedly present new challenges, new obstacles to overcome. But Leah faced them now with a newfound confidence, a quiet strength born from the victory at the center fence. Deep within her heart, she knew she could face any challenge and conquer any fear if she had Whisper by her side and held to the unwavering belief in her ability. This wasn't the end of a competition; it was the beginning of a new chapter filled with the promise of future triumphs, built upon the foundation of the hard-won victory at the heart-stopping center fence.

The memory of that flight, that moment of perfect synchronization, the feeling of total mastery, would fuel her ambition, guide her choices, and keep her centered as she moved forward, not a competitive rider, but a strong, confident young woman, forever changed by the experience. The lessons learned were far more significant than any prize ribbon. She understood that genuine victory lay not in the accomplishment but in the strength found within oneself and the enduring partnership with the magnificent creature that carried her to it.

# Chapter 13: Success or Failure

The moment extended, time itself seeming to slow. Whisper, my magnificent mare, met the fence not with a hesitant leap, but with a surge of power that belied her graceful form. The power wasn't brute force; I felt the powerful thrust of her hindquarters, the effortless lift as we rose, defying gravity for a fleeting moment.

Up, up we soared, the ground falling away beneath us. After a moment's pause, the world became a vibrant blur. The crowd was a silent hum, a background noise to my heart pounding. All I could focus on was Whisper, her rhythmic breathing, the feel of her warm body against mine, the absolute trust that bound us together. The height of the fence was immense, a towering obstacle that had filled me with dread moments ago. However, it was a momentary challenge, a hurdle to overcome.

Then, the descent. It was as smooth as the ascent, a controlled drop that barely disturbed our rhythm. Whisper landed with a barely perceptible shift, four hooves planting firmly on the ground. There was no stumble, no hesitation, a silent acknowledgment of a job well done, the effortless grace of a perfectly executed jump. A wave of relief, so intense it was almost physical, washed over me. We had cleared the center fence.

The silence that followed was deafening, broken only by the pounding of my heart as it slowly returned to a normal rhythm. Then, the crowd gasped and cheered wildly, applause shaking the ground. I didn't fully hear the overwhelming sound. The rest of the course felt like a dream. Once imposing challenges, the remaining jumps now seemed almost insignificant. My confidence soared, boosted by the victory at the center fence. Sensing my change in demeanor, Whisper responded with an even greater sense of cooperation. We moved as one, a seamless blend of horse and rider, our movements fluid and precise. Every jump replicated the perfection I had achieved at the center fence, with smooth and controlled landings.

As we crossed the finish line, the crowd's roar was deafening, a wave of sound engulfing me. But amidst the cheers, I heard Whisper's breath, steady, a gentle rhythm that grounded me. This wasn't a victory or testament to our partnership, trust, and unwavering bond. Afterwards, ribbons and awards seemed trivial, poorly representing the event's importance. The awards did not define the genuine triumph received, but by the profound sense of achievement, the experience of pushing past my boundaries, and the unbreakable connection I forged with Whisper. Although the accolades were a pleasant reward and validated the hard work and dedication we invested, the internal triumph that swelled within me far surpassed any external recognition.

The euphoria of the win lasted for days. But beyond the initial excitement, a deeper understanding took root. This wasn't about winning a competition but overcoming a deep-seated fear, confronting my self-doubt, and emerging victorious. Once a symbol of my anxieties, the center fence now represented my personal growth, a testament to my resilience and determination. The memory of that jump—the surge of power, the moment of suspended flight, the feeling of perfect synchronization between horse and rider—would stay with me forever. It symbolized my potential, a potent reminder of what I could achieve when I pushed beyond my self-imposed limits.

Its impact echoed through my entire existence in the ensuing period. My confidence blossomed, transforming my approach to training, interactions with fellow competitors, and the overall outlook on life. I approached each training session with renewed vigor, pushing myself and Whisper to reach new heights. The shadow of self-doubt that lingered over me retreated, replaced by a quiet assurance in my abilities. My relationship with Whisper deepened, and our shared experience of overcoming a significant obstacle strengthened our bond. Our communication went beyond simple words; instead, it reached a more profound understanding that blossomed and expanded with every training session and competition we shared. It wasn't simply a horse and rider; our connection was more profound, stronger, exceeding the sum of its parts.

I approached the subsequent competition with increased confidence. The other fences, once daunting obstacles, were now mere stepping stones on my path to victory. I rode with an unshakable calmness, a steady hand on the reins, and my focus unwavering. The success I experienced in subsequent competitions confirmed my growth and validated my transformation. Each victory built upon the foundation of the center fence jump, a testament to my growing skill and unwavering belief in myself.

This wasn't about equestrian sports; it was about life itself. The lessons learned, extended far beyond the confines of the show jumping arena. I learned to overcome my fears, trust my abilities, and rely on my partner's unwavering support. The center fence jump wasn't simply a jump, but a metaphor for life's challenges, an embodiment of resilience, determination, and the power of self-belief.

The journey wasn't over, of course. Increased competition causes surmounting greater obstacles. However, I approached the future with a different perspective and strength. With a certainty that resonated deep within my soul, I knew I could face any obstacle and conquer any fear, as long as I had Whisper by my side and as long as I held to the unshakeable belief in my potential. Once a symbol of my fears, the center fence had transformed into a symbol of my triumph — a beacon of hope guiding me toward a future filled with boundless possibilities. And that was a victory far greater than any prize I could ever win. The memory of that flight, that moment of perfect synchronization, would fuel my ambition and inspire my journey.

# Chapter 14: Aftermath of the Jump

The world spun, a dizzying blur of color and motion, before settling into a slow, almost painful stillness. My breath hitched in my throat, a ragged gasp that felt strangely out of place in the sudden quiet. The roar of the crowd, a moment ago a deafening wave, had receded to a distant murmur, replaced by the frantic thump-thump-thump of my heart against my ribs. My muscles ached, a pleasant soreness that spoke of exertion and the release of adrenaline. My magnificent mare, Whisper, remained perfectly still, her flanks rising and falling gently, a testament to her incredible power and athleticism during that impressive jump. She remained completely steady and did not tremble at all. As she stood there, the gentleness in her eyes, so soft and calm, perfectly reflected the profound sense of relief that washed over me, a feeling so profound and overwhelming it filled my entire being.

With my hands still tightly gripping the reins, a numbing sensation washed over me, yet the smooth leather felt strangely comforting against my trembling fingers. A wave of intense vulnerability unexpectedly overturned my expectation of a powerful, triumphant feeling and an explosion of pure joy, leaving me exposed and fragile. It wasn't exactly fear, but it was a feeling that was very similar to fear. It was an experience far exceeding the commonplace, a visceral, unfiltered emotion that stole my breath and left me feeling profoundly unsteady. Having pushed myself to the absolute limit, I found myself in the aftermath, grappling with a fear so profound that it had threatened to paralyze me, leaving me shaken and vulnerable. My survival and victory left me reeling, completely overwhelmed by the sheer force and intensity of my experience.

Tears welled in my eyes, hot and stinging, a silent testament to the emotional turmoil that raged within. They were tears not of defeat but relief, exhaustion, and a profound and unexpected vulnerability. It was a release, a letting go of the tension that had coiled within me for days, weeks, even months. The weight of expectation, the pressure to live up to my family's legacy, and the crushing weight of self-doubt seemed to dissipate after that breathtaking jump.

Slowly, I released the reins, my fingers loosening their grip.

Whisper shifted slightly, her body relaxing as she sensed my change in posture. I slid a hand down her neck, feeling the warmth of her skin beneath my fingertips. Her calm and even breath was a grounding presence, a silent reassurance that we had done it. We had faced the center fence, the dreaded obstacle that had haunted my nightmares, and emerged victorious.

The silence stretched, a heavy blanket that enveloped us, broken only by the occasional shuffling of feet in the crowd. Then, a ripple of applause, growing louder, building into a thunderous ovation that vibrated through the ground. It washed over me, a wave of sound that finally broke through the emotional numbness. I looked up, my eyes searching for my trainer, Mr. Henderson, his face a mixture of pride and relief. He raised his hand, a triumphant gesture that acknowledged not the flawless jump, but also the immense courage it had taken to execute it.

My father, his usual stern demeanor softened, was clapping enthusiastically, his eyes shining with something akin to pride.

Even Emma, my rival, who usually wears a mask of cool indifference, offered a grudging nod of acknowledgement. It was a moment of shared recognition, a silent understanding that transcended the boundaries of competition. We had all witnessed something extraordinary beyond the mere clearing of a fence.

As I dismounted, my legs were weak, but my heart soared with a sense of accomplishment that had nothing to do with ribbons or awards. It was a triumph of spirit, a victory over my self-doubt. Whisper nuzzled my hand, her soft breath a quiet reminder of the bond we shared.

A flurry of polite exchanges and carefully constructed smiles characterized the awards' presentation, leaving an indistinct and fleeting impression on those present. Faced with my overwhelming fulfillment, the ribbons I wore felt insignificant, like a mere accessory of little consequence. The accolades we received served as a gratifying and welcome confirmation of the success we had achieved. However, these external accolades were insignificant compared to my profound internal sense of accomplishment, a deep knowing that I had bravely confronted my fears and emerged from the experience a more assertive, self-assured, and confident individual.

A whirlwind of congratulations, many interviews, and intense media attention characterized the days immediately following the event. My victory at the center fence had unexpectedly thrust me into the public eye, and I was now struggling to adjust to this sudden and significant increase in public attention and recognition. The pressure was intense, yet somehow different and intensified from what I had felt before. The feeling wasn't the heavy, crushing weight of self-doubt that bore down on him, but something far more insidious.

However, the quiet moments held the most tremendous significance amidst the excitement. The quiet moments spent grooming Whisper, the gentle touch of her velvety nose against my cheek, and the shared understanding passed between us without words. Those were the moments that truly mattered, the moments that solidified the bond we had forged, the moments that reminded me of the profound transformation that had taken place within me.

That center fence, which had once represented a powerful symbol of my deepest fears and anxieties, now stood as a powerful testament to my remarkable resilience and ability to overcome challenges. As I looked at the fence, I realized how perseverance, unwavering dedication from a true partner, and their support are crucial in overcoming even the most daunting obstacles and most profound feelings of self-doubt. Although the journey was arduous, full of challenges and obstacles that tested our limits, the reward we received was immeasurable, surpassing all expectations and making the hardships worthwhile. Winning wasn't the primary aim; other, more important factors were at play. More than any material reward or accolade, that achievement represented a triumph deserving of significant celebration and recognition.

# Chapter 15: Reflection and Growth

As I guided Whisper back to her stall, the cool night air gently caressed my skin, a pleasant contrast to the warmth of the day. From the quiet intimacy of the stable block, where a peaceful calm reigned, the sounds of the bustling showgrounds, including the excited chatter and celebratory whoops of the attendees, seemed distant and muffled, as if separated by a thick, sound-dampening barrier. As the adrenaline wore off, a satisfying tiredness filled my body and settled comfortably within my very bones. As if seared into his memory, the jump stayed vividly sharp, precise and clear in his mind, with the sharpness of a brand new blade.

It wasn't the physical exertion, the strain on my muscles, and the intense focus required to navigate Whisper through that demanding course that left the most profound impression. The emotional journey, the internal battle I'd waged against my self-doubt, left the most profound impression. The center fence was not an obstacle; it reflected my fears and the strength to overcome them.

I remembered the weeks leading up to the competition, the gnawing anxiety that had accompanied every training session. The pressure, the relentless weight of expectation, had been almost unbearable. My family's legacy in equestrian sports was a double-edged sword. While it fueled my ambition, it also instilled a crippling fear of failure. I couldn't bear the thought of disappointing them, of falling short of their high standards. The idea of letting myself down, of proving my self-doubt correct, was almost more terrifying than the prospect of losing.

With her effortless grace and unwavering confidence, Emma only intensified my anxieties. Her nonchalant air of superiority and seemingly effortless victories had become a constant source of comparison, a nagging reminder of my shortcomings. I often scrutinized her every move, dissected her technique, and searched for flaws where none seemed to exist. This internal competition had become as intense, if not more so, than the competition itself.

But facing the center fence, that imposing obstacle that had loomed large in my imagination for weeks, had changed everything. In that moment, suspended between the earth and the sky, I transcended my anxieties with Whisper's powerful muscles propelling us forward. It wasn't a sudden, magical transformation, but a gradual process culminating in a powerful realization. I discovered a resilience within myself, a strength I hadn't known I possessed.

It wasn't merely about clearing the fence; it was about removing the mental hurdles, the self-imposed limitations I'd placed upon myself. The physical act of jumping merely manifested the internal struggle I had overcome. Whisper, my steadfast partner, had been instrumental in this journey. Her unwavering trust and quiet strength had been a constant source of reassurance, a silent affirmation of my capabilities. We were a team, a partnership forged in shared dedication, unwavering trust, and mutual respect.

The celebratory dinner that followed the competition was a blur of polite conversation and forced smiles. Yet, even amidst the chaos, I retreated into moments of quiet introspection. Upon reviewing the lessons learned, personal development defined success and overcoming physical and emotional challenges, rather than by receiving ribbons or trophies. It was about discovering a strength and resilience I never knew I possessed.

The following days were filled with media interviews and public appearances. The sudden attention was overwhelming, yet I discovered a new confidence, a readiness to share my story and inspire others to pursue their passions, even in the face of fear and self-doubt. The spotlight, initially daunting, became a platform for sharing my message of perseverance and the power of self-belief. I learned to embrace the challenges that arose, seeing them not as obstacles, but as opportunities for continued growth.

My relationship with Emma also underwent a subtle shift. Despite our rivalry, we developed a grudging respect for each other. We no longer engaged in the silent, competitive staring matches that had marked our previous encounters. Our interactions became more civil, tinged with a mutual understanding of the dedication, sacrifice, and hard work required to achieve excellence in our chosen sport.

The center fence remained etched in my memory, not as a symbol of fear or self-doubt, but as a marker of a profound transformation. It was a potent reminder of the power of perseverance and the importance of believing in oneself, even in the face of seemingly insurmountable challenges. It taught me that the accurate measure of success lies not in the absence of fear, but in the courage to face it head-on and emerge victorious.

The journey was far from over. New challenges and hurdles would undoubtedly arise, but I approached them with a newfound confidence and a strengthened sense of self-belief. I had learned to rely on my instincts, trust my abilities, and draw strength from the unwavering support of those around me. The experience with the center fence had not changed my perspective on show jumping. Still, it had reshaped my understanding of myself, my capabilities, and my place in the equestrian world.

The days turned into weeks, and the weeks into months. The initial euphoria of victory subsided, replaced by a quiet contentment — a deep-seated satisfaction that transcended the accolades and awards. I appreciated the simple things—the feel of Whisper's soft coat beneath my hands, the comforting rhythm of her breath as she slept peacefully in her stall, and the quiet moments spent grooming her, cleaning her tack, and preparing for the next challenge.

The training continued, each session a testament to the progress I'd made, both physically and mentally. The fear remained, of course, but it was no longer a paralyzing force. It was a familiar companion, a subtle reminder of the challenges I'd faced and overcome. I embraced it, recognizing it as a part of the journey, a catalyst for growth and improvement.

The upcoming competitions felt different. The pressure was still there, but it no longer suffocated me. I approached each jump with a newfound clarity, a calmness that emanated from within. I was more attuned to Whisper's responses, more intuitive in my riding, more confident in my abilities. My focus shifted from the outcome to the process itself, from the potential for failure to the joy of the ride.

My relationships with my family had also deepened. The shared experience of overcoming adversity, the mutual understanding forged in the crucible of competition, created a stronger bond than any mere victory could have achieved. We celebrated not the ribbons and the trophies, but the journey itself, the growth, and the transformation that had taken place. The legacy I sought to uphold was no longer a source of overwhelming pressure; it was a source of inspiration, a testament to the enduring power of family, passion, and dedication.

In retrospect, the center fence represented far more than just a hurdle to be jumped. In a powerful display of symbolism, the sculpture captured the essence of life's many trials—the challenges, obstacles, and tests of resilience—while also highlighting the insidious and often debilitating presence of fear. With considerable power, the film showcased the incredible ability of the human spirit to not only conquer significant challenges and rise above fear, but also to emerge from the experience with enhanced strength and resilience. We found the journey to be incredibly arduous, demanding, and agonizing, a truly grueling experience that severely tested our limits of endurance and resilience, pushing us to our very edge.

The victory was not solely my accomplishment, but a collective triumph achieved through the efforts and support of many. And that, as I reflected upon it, proved to be the most genuine and significant triumph of my entire experience. I constantly draw inspiration from the memory of that jump, a moment of intense vulnerability followed by exhilarating triumph, reminding myself that overcoming the greatest challenges often yields the most significant personal growth and development.

# Chapter 16: Navigating the Remaining Course

The next jump, a triple bar, felt different. Before, the thought of a triple bar would have sent a shiver down my spine, a surge of anxiety tightening my chest. Now, as I approached it, a calmness settled over me. It wasn't the absence of fear, but a conscious decision to acknowledge it, accept it as part of the process, and then move beyond it. Whisper sensed the shift in my demeanor, her powerful strides reflecting my newfound composure. We cleared the triple bar with effortless grace. The rhythmic pounding of her hooves a steady beat against the tense silence of the arena.

The following obstacle, a vast and imposing one, tested my precision and timing. My focus sharpened, my attention homed in on the exact take-off point, the precise moment of release. It was a dance, a partnership between horse and rider, a silent conversation conducted through subtle shifts in weight and pressure. There was no room for hesitation, no space for self-doubt. Only the unwavering trust between Whisper and me guided our movements, leading us to a flawless clearance.

The center fence loomed before us, which was always a source of apprehension for even the most seasoned riders. The gleaming surface reflected the anxious faces in the stands, their hushed whispers a low hum in the background. I could feel the weight of expectations and the pressure to perform, but it was different now. This was no longer a threat, a challenge to overcome, but a test, an opportunity to show my progress and the strength I'd unearthed. Whisper, sensing my calmness, approached the fence with a quiet confidence that reassured me. We flew over the jumps in a seamless, graceful arc, a testament to our shared journey.

The last line of jumps, a series of tightly spaced verticals and obstacles, demanded exceptional precision and control. Each jump was a challenge, a test of focus and skill. But with each clean jump, my confidence soared. The fear, though still present, had become a manageable companion, a constant reminder of my past struggles, yet also a source of quiet strength. It motivated me to excel. The course seemed to shrink with each successful jump, the obstacles morphing from daunting impediments to mere stepping stones on the path to victory.

The euphoria was almost palpable when we reached the final fence, a towering obstacle decorated with lush greenery. It wasn't the potential for victory, but the triumphant feeling of overcoming my self-doubt and the exhilaration of having pushed beyond my limitations. It was the culmination of weeks of training, months of relentless self-improvement. As we cleared the final fence, a roar erupted from the stands, a wave of sound that washed over me, lifting me and carrying me aloft on a tide of relief, pride, and overwhelming joy.

The following moments were a blur. The judges' scores, the crowd cheers, and congratulatory hugs from my family and trainer blended into a symphony of joy and relief. But amid the celebration, I retreated to a quiet corner, seeking solitude to reflect on the journey that had brought me there. The victory felt different this time, more profound, more meaningful. It was less about the ribbons and the trophies, and more about the transformation I'd undergone, the self-discovery I'd made. The center fence, the triple bar, the jump, the last obstacle—each jump was a milestone on my self-belief journey. Later that evening, as I groomed Whisper, I felt a profound sense of gratitude. Her unwavering trust and strength had been instrumental in my success. She wasn't a horse; she was my partner, confidante, and friend. Our connection ran deeper than mere competition; it was a bond forged in mutual respect, shared challenges, and unwavering dedication. As I brushed her soft coat, I felt renewed determination to continue our journey and face whatever challenges lay ahead.

The following days brought a wave of media attention. The media widely shared interviews, articles, photographs, and my story. It wasn't about fame or recognition; it was about the opportunity to inspire others and share my message of perseverance and self-belief. I spoke openly about my struggles, anxieties, and journey towards self-discovery. I shared my story not as a boast of victory, but as a testament to the human spirit's capacity to overcome challenges, to rise above adversity, and to emerge stronger and more resilient.

My relationship with Emma also shifted. While the competition remained, a grudging respect blossomed between us. We acknowledged each other's skill, dedication, and unwavering pursuit of excellence. A quiet understanding, a mutual appreciation for the shared challenges, and a mutual respect for our achievements replaced the silent staring matches and the tense atmosphere that had permeated our previous encounters.

The rest of the competition season passed in a blur of training, competitions, and quiet moments of reflection. Each competition was a step forward, a further demonstration of the growth I'd experienced. My confidence grew with each successful round, the fear remaining, but no longer paralyzing, instead catalyzing greater focus and intensity. Working in partnership with Whisper and the joy of the ride became my driving force, overshadowing competitive pressures.

The series culminated in the national championships, a high-stakes competition that attracted the nation's most skilled and accomplished riders. Despite the incredibly intense pressure and fierce competition, I felt well-prepared for whatever challenges might come my way. Having confronted and overcome my previous fears and self-doubt, I emerged from the experience a stronger and more confident individual. Precision, intense focus, and an effortless grace that made it a truly memorable experience marked my riding. Whisper's response was magnificent in its accuracy, her movements a perfect reflection of my calm and determined advance, as if she were a mirror image responding to my every subtle gesture.

The victory was not rightfully mine; it belonged to someone else. The journey proved to arduous, demanding, and excruciatingly painful, testing the limits of both physical and mental endurance. The challenges, while significant, had been far surpassed by the magnitude of the rewards reaped. My transformation as both a rider and a person was so profound and lasting that it changed me in ways I never expected. In my memory, the center fence stands not as a symbol of fear, but as a powerful testament to perseverance, highlighting the crucial role of self-belief and underscoring the remarkable, enduring strength of the human spirit. The experience served as a potent reminder that a life devoid of fear did not define genuine success, but by the unwavering courage to confront fear directly and ultimately triumph over it, emerging stronger and more resilient.

# Chapter 17: Emma's Performance

The atmosphere crackled with anticipation as Emma entered the arena. Her chestnut mare, a fiery creature named Ember, practically vibrating with energy. Emma, unlike Leah, exuded an aura of effortless confidence — a self-assurance that bordered on arrogance. She sat tall in the saddle, her posture impeccable, her movements fluid and precise. There was a sharpness to her gaze, a steely determination that hinted at the fiercely competitive spirit simmering beneath the surface.

Leah observed with mixed feelings of admiration and apprehension. Emma's riding style was the antithesis of Leah's. While quiet intensity and a focus that stemmed from a deep connection with Whisper characterized Leah's approach, Emma's style was flamboyant, almost theatrical. She seemed to ride not the course, but the crowd, her every movement calculated to impress, to captivate.

Ember, a mirror image of her rider, responded with the same breathtaking precision and explosive power. They tackled the jumps with a dazzling showmanship that had the crowd roaring their approval. The jumps that had tested Leah's nerve, causing moments of doubt and anxiety, seemed to be mere stepping stones for Emma and Ember. They cleared every obstacle with effortless grace, their combined movements a symphony of skill and confidence. It was breathtaking to watch, a captivating display of equestrian prowess.

There was a controlled aggression in Emma's riding that Leah couldn't help but admire. It differed from her approach, but was no less effective. Where Leah sought harmony and connection with Whisper, finding strength in their shared trust, Emma seemed to command Ember's power, pushing the mare to her limits, demanding the ultimate performance. The two appeared locked in a silent duel, a nonverbal exchange of skill and ambition played out against the backdrop of the competition. While Leah focused on a quiet determination, Emma projected an aura of unshakeable confidence, almost to the point of nonchalance.

As Emma navigated the intricate course, the similarities and differences between her performance and Leah's became starkly apparent. Both riders displayed exceptional skill and mastery, yet their approaches diverged dramatically. Emma's style was one of audacious power, a display of breathtaking athleticism and calculated risk-taking. She rode with a flair that captivated the audience, while Leah's performance was a study of controlled precision and quiet intensity. Leah's connection with Whisper spoke of a profound partnership, a bond forged in mutual trust and respect. Emma's connection with Ember was more forceful, a display of dominance and control. This made for an intriguing contrast, highlighting the different ways in which riders and their horses could achieve the same goal—impeccable performance in a demanding competition.

The tension between the two riders, previously simmering beneath the surface, became palpable as the competition progressed. It wasn't overt hostility, but a silent rivalry, a constant, underlying current of competitiveness that permeated their every move. Each clean jump by Emma was a subtle challenge to Leah, a reminder of the high standards they both aspired to. But Leah, hardened by her struggles and strengthened by her victories, met the challenge with unwavering determination. She used her victories not as a cause for complacency, but as further motivation to improve, push herself further, and continue refining her partnership with Whisper.

The contrasting riding styles and competitive spirits created a powerful dynamic, capturing the attention of both judges and spectators. It was a captivating rivalry, not only for its intense competitiveness but also for the distinct riding styles it showcased. Both riders were exceptionally talented, but their approaches differed significantly. Each possessed unique strengths, highlighting the diversity of equestrian skill and athleticism.

The final round of the competition proved to be a nail-biting affair.

Maintaining her flamboyant style, Emma pushed Ember to the absolute limit, delivering a thrilling and risky performance. Leah, in contrast, rode with a measured intensity, her focus unwavering. It was a captivating showdown, featuring two distinct riding styles and two powerful horses, each striving for perfection. The judges, impressed by both riders, deliberated for a considerable time before announcing the results.

The announcement of the scores created a breathless atmosphere. Emma had edged out Leah with her daring style by a fraction of a point. The crowd erupted, cheers and applause washing over the arena. Leah, though disappointed, felt a grudging admiration for Emma's performance. It had been a worthy contest, a battle between two distinct equestrian philosophies, two incredibly talented young women. The rivalry, while fierce, was born out of respect, a recognition of each other's skills and determination.

Later, as Leah groomed Whisper, she reflected on Emma's performance and the intensity of their competition. Though she had longed for victory, she acknowledged the talent and skill her rival had shown. The experience had pushed her beyond what she thought possible, honing her abilities and strengthening her resolve. This rivalry, born from the high-stakes world of competitive show jumping, pushed her forward, enhancing her skill and inspiring her growth as an equestrian.

In the following weeks, the rivalry continued, but it took on a new, more mature dimension. Emma's audacious riding style and almost brash confidence contrasted with Leah's quiet determination and intense focus, but each learned to appreciate the other's strengths. Their competitive spirit remained, but mutual respect tempered it; they tacitly acknowledged each other's capabilities, forming a silent pact born of their shared passion for equestrian sports. The rivalry between them acted as a catalyst, not only driving both riders to achieve greater heights but also refining their skills and pushing them to explore the boundaries of their potential, resulting in significant improvements. No longer focused on asserting dominance or superiority, the rivalry developed into a dynamic exchange where participants spurred each other on, relentlessly pursuing perfection and experiencing the exhilarating heights of peak competition. As the journey continued onward, it set the stage for a more formidable challenge that lay ahead in the upcoming phase of the competition season. Although their relationship began in the heat of fierce competition, they forged a unique bond, not through friendship, but through mutual admiration born from their shared challenges.

# Chapter 18: Unexpected Setbacks

The following morning dawned bright and crisp, promising a perfect day for show jumping. Leah felt a familiar knot of anxiety tighten in her stomach as she led Whisper to the warm-up area. Whisper, usually so eager, seemed subdued, her movements less fluid than usual. Leah's meticulous pre-competition routine—checking tack, brushing Whisper's coat, and mentally rehearsing the course—felt almost mechanical, her mind preoccupied with a nagging unease. She'd noticed a slight stiffness in Whisper's hindquarters the previous evening, a subtle lameness that she'd initially dismissed as fatigue. The stiffness persisted, and a chilling thought struck her: Whisper might be injured.

Panic threatened to overwhelm her. This was the most crucial competition of the season, which could solidify her position as a rising star in the equestrian world—even with a potentially injured horse. Withdrawing now felt like utter defeat. But the thought of riding Whisper while she was in pain was unbearable. She ran a hand over Whisper's flank, feeling the subtle tension in her muscles. The usually soft, yielding flesh felt taut and rigid.

Leah sought her trainer, Mr. Henderson, a seasoned equestrian with decades of experience. He scrutinized Whisper, his experienced eyes scrutinizing every movement and subtle shift in posture. His expression was grave. "She's not right," he murmured, his voice low and serious. "We need to get the vet."

The veterinarian arrived promptly, his arrival adding to the mounting tension. After a thorough examination, he confirmed Leah's fears. Whisper had sustained a minor strain on her hindquarters, likely from the intense exertion of the previous day's competition. Although not a severe injury, it seriously affected her performance. Riding her would risk further injury, potentially ending her competitive season. The news hit Leah like a physical blow. All her hard work, countless hours of training, and unwavering dedication threatened to crumble instantly. The image of Emma, her confident rival, flashed through her mind, her victory now even more distant and unattainable. Tears threatened to spill, but Leah clenched her jaw, determined to maintain her composure.

Noticing the profound despair etched upon her face, Mr. Henderson, in a gesture of comfort and reassurance, gently placed his hand on her shoulder. With a gentle tone, he addressed Leah, explaining that she shouldn't feel any shame or embarrassment about choosing to withdraw.

Although his words should comfort her and soothe her distress, they ultimately failed to ease her profound disappointment. In her mind's eye, she pictured this competition not merely as a contest, but as her pivotal moment of triumph, a chance definitively to prove her capabilities and conquer the persistent nagging voice of self-doubt that had held her back. Cruelty and unexpectedness defined the ripping away of my expected chance. The withdrawal was an agonizing decision, but it was necessary.

The news spread quickly through the equestrian community, eliciting sympathy and speculation. Some whispered that Leah was not up to the challenge and had cracked under pressure. Others sympathized, understanding the sacrifices and dedication involved in competitive equestrian sports. Emma approached Leah with an empathetic demeanor, a contrast to her usual behavior.

"I heard about Whisper," Emma said, her voice surprisingly soft. "That's a real shame."

Emma's unexpected kindness took Leah aback. "Thanks," she replied, her voice barely a whisper.

"I know how much this means to you," Emma continued. "It's a tough break. I wouldn't wish this on my worst enemy."

The genuine concern in Emma's voice was surprising. It wasn't the same Emma who exuded a constant aura of self-assurance and competitive drive. This was a different side of her, a vulnerability Leah had never witnessed before. She spent the next few days nursing Whisper back to health. Leah dedicated herself to Whisper's recovery, spending hours with her, applying gentle liniments, and whispering to her. The quiet time allowed her to reflect on her experiences and confront her self-doubt. She recognized that her expectations and ambition had perhaps become too demanding and relentless. She'd pushed both herself and Whisper to their limits, ignoring the subtle warning signs of strain and fatigue.

The setback, painful as it was, served as a valuable lesson. It taught her the importance of balance, of recognizing the limits of both horse and rider, of nurturing the bond of partnership rather than pushing it to the breaking point. She learned that genuine success wasn't about winning; it was about the enduring connection with her horse, the commitment to their shared well-being. The enforced break allowed Leah to reassess her training regimen. She collaborated with Mr. Henderson to develop a more comprehensive approach, prioritizing Whisper's physical and emotional well-being. They implemented new techniques, focusing on flexibility and preventive care, which reduced the risk of future injuries.

The recovery period also provided time for self-reflection. Leah confronted her tendency towards self-criticism and perfectionism, acknowledging that, while ambition was crucial, it needed to be tempered with self-compassion and a realistic understanding of her limitations. She learned to listen to Whisper's subtle cues, recognizing that their partnership required mutual respect and sensitivity. As Whisper's recovery progressed, so did Leah's emotional healing. She found solace in the quiet moments spent with her horse, reinforcing their deep bond. The incident had shaken her, but it had also strengthened her resolve and deepened her understanding of the intricate dance between rider and horse.

The following competition was not as significant as the one she'd missed, but it was a crucial stepping stone. Leah and Whisper entered the arena, not with the same pressure and expectation as before, but with a newfound perspective. They focused on the joy of the sport, the thrill of the challenge, and their unique partnership. While not a stunning victory, their performance was solid and respectable, showcasing resilience and progress. The unexpected setback had ultimately strengthened their bond and refined their approach to competition. Leah learned a valuable lesson from the experience: success isn't just victory, but the journey, growth, and bond between horse and rider. She found a deeper, more profound satisfaction than any trophy could ever provide. The journey continued, with Leah and Whisper, now stronger and wiser, ready to face the challenges that the next competition would bring.

# Chapter 19: Support System

Although Leah was determined, the lingering pain of withdrawal still caused a dull ache beneath her renewed resolve. Yet, in the wake of her disappointment, a quiet strength emerged, steadily growing and fueled by the constant and unwavering support of those who cared for her. Providing unwavering support and a deep well of comfort and understanding, her family, the constant bedrock of her life, rallied around her during this difficult time.

Her father, a former champion show jumper, knew the weight of expectation and the sting of setbacks. He shared his own experiences with injuries, missed opportunities, and the resilience needed to rise again. His words reassured her that her feelings were valid and her desire to return to competition justified.

Her mother, while less familiar with the intricacies of show jumping, provided a different support. Her quiet strength, unwavering belief in Leah's abilities, and the practical help of caring for Whisper were invaluable. She brought Leah comforting meals, ensuring she was well-fed and rested, and subtly managed the household chores, allowing Leah to focus her energy on Whisper's recovery. She listened patiently to Leah's anxieties, offering gentle encouragement and a calming presence that soothed Leah's turbulent emotions. Their conversations were rare about equestrian sports. Instead, they revolved around shared laughter, comforting silences, and simple acts of affection—a warm hug, a shared cup of tea, or a quiet evening watching a movie. These slight gestures were incredibly potent, nourishing Leah's spirit and bolstering her confidence.

Besides the comfort she received from her immediate family, Leah discovered unexpected solace and support from her circle of friends, a fact that surprised and comforted her. Sarah, who was not only her closest friend but also a fellow rider possessing a keen understanding of the equestrian world, made it a point to visit every day. When Leah felt overwhelmed, Sarah's calm and reassuring presence helped to stabilize her and restore a sense of calm. During Sarah's recovery, Sarah's practical help, which included assisting with Whisper's care and keeping her informed about the happenings in their equestrian community, helped to prevent her from feeling isolated. Leah consistently drew strength from her friend's empathetic ear and unwavering belief in her capacity to triumph over the setback.

Then there was Mark, a quieter friend, but equally supportive of his unique way. Mark's technical expertise, more than his equestrian skills, provided a different perspective that often helped Leah solve Whisper's rehabilitation problems. Mark's perspective wasn't clouded by competitive rivalry; he offered solutions rooted in a deep understanding of animal care and physiotherapy. His calm demeanor and technical advice proved invaluable as he helped Leah devise a tailor-made rehabilitation plan for Whisper.

Mr. Henderson, her trainer, played a pivotal role in her recovery.

He was not simply a coach, but a mentor, a friend, and a father figure to his players. Recognizing the profound nature of Leah's disappointment, he responded with both empathetic understanding and the provision of firm, helpful guidance. Although he always respected her feelings and never made light of them, he also stressed the importance of maintaining perspective and learning from any difficulties she faced. His method involved a combination of tough love, which pushed his subjects to grow, and unwavering support that provided a safety net as they struggled to improve. He was a constant presence, his presence a quiet reassurance, both during Leah's moments of despair and during the hours she dedicated himself to Whisper's rehabilitation. He remained committed to her, tailoring a revised training program that addressed both horse and rider's physical and emotional needs, ensuring a slower, more sustainable path to recovery. His patience, experience, and unwavering belief in her abilities helped Leah navigate this challenging period, guiding her towards a balanced and holistic approach to equestrian sports. He pushed her to improve, not in terms of competition results, but her horsemanship, her understanding of Whisper, and their synergy.

The veterinarian, Dr. Evans, offered more than medical expertise. Dr. Evans regularly updated Leah on Whisper's progress, explaining the healing process so she could easily understand. He answered her many questions patiently, reassuring her and explaining that Whisper's recovery was progressing well. He emphasized the importance of steady rehabilitation and highlighted potential long-term strategies to prevent future strain. His calm demeanor and reassuring words helped ease Leah's anxiety, allowing her to trust the process and focus on ensuring a complete recovery for Whisper. He didn't treat an injury; he addressed the emotional toll the situation took on Leah, understanding the emotional bond between human and animal, a connection that ran deeper than mere athlete and horse.

The support network surrounding Leah extended beyond these core individuals. Fellow competitors, initially fueled by speculation, soon showed genuine empathy. Some offered words of encouragement, sharing their experiences with setbacks and injuries. Others provided practical help, sharing tips on rehabilitation techniques or recommending equine therapists. Initially marked by rivalry, the community now showcased its solidarity, a supportive collective eager to assist Leah in her recovery process.

Even Emma, her competitive rival, continued to show an unexpected empathy. She remained in touch, offering advice and words of encouragement. While not overly effusive, her support was significant, silently acknowledging Leah's talent and resilience. This unexpected support removed a layer of isolating pressure, allowing Leah to focus on her recovery process. Emma's support was unsurprising; This unexpected alliance forged a bond based on mutual respect and a shared passion for equestrian sports, showing that competition didn't have to negate friendship or empathy.

The combination of this broad and diverse support network was pivotal to Leah's recovery. It was the belief in her, knowing she wasn't alone, and understanding that setbacks are part of the journey. The support she received from her family, trainer, friends, and the wider equestrian community helped her navigate this challenging period, fostering resilience, growth, and a deeper understanding of what it truly meant to be a successful equestrian. A strong support system's vital role in high-stakes competitions became clear, demonstrating that success depends not only on individual talent but also on the collaborative efforts of a supportive network. While initially devastating, the enforced break ultimately led to a more profound understanding of herself, her horse, and the importance of community, solidifying her journey towards becoming a stronger, more resilient, and ultimately more successful equestrian.

The support system was not simply a comforting presence, but the foundation upon which she rebuilt her hopes, dreams, and career.

# Chapter 20: Building Resilience

The next few days were a blur of controlled chaos. Whisper's near-withdrawal led to a concentrated effort. They adjusted Leah's riding schedule, but did not abandon it. Mr. Henderson, ever the pragmatist, implemented a changed training regime, focusing on exercises that built strength and confidence without pushing Whisper too hard. It wasn't about winning anymore; it was about rebuilding trust, strengthening the bond between horse and rider, and solidifying Leah's mental fortitude. Each session was a testament to the power of patience and perseverance. Instead of the high-pressure atmosphere of intense competition practice, a quiet intensity marked their sessions, focused on the subtle nuances of communication and cooperation between horse and rider. The pressure was internal, self-imposed, a desire to prove she could overcome this setback.

During the competition, there was a subtle, almost imperceptible shift in the atmosphere. As time went on, the initial period of gossip and speculation gradually subsided, replaced by a quiet and respectful atmosphere. As Leah completed, she registered the subtle nods of acknowledgment and the silent words of encouragement offered by competitors she had never previously met, a silent camaraderie that surprised and heartened her. Although the competitive spirit remained a driving force, a strong sense of camaraderie and mutual understanding of the inherent risks and vulnerabilities that were a part of their chosen athletic pursuit notably softened and balanced it. They developed a newfound appreciation for the intricate and delicate balance between ambitious aspirations and the unwavering resilience needed to overcome inevitable obstacles. The quiet strength Leah showed subtly revealed a vulnerability that the other riders, recognizing from their shared experiences, acknowledged with a mutual respect born of their bond.

One morning, Sarah excitedly approached Leah in the sunlit stables. "I saw Emma warming up. She's amazing, as usual. But something feels...different. More focused. Maybe even... nervous?" Leah smiled, a small, knowing smile. She understood. Emma, with all her outward confidence, was also feeling the pressure. There is pressure to maintain her reputation and consistently deliver top-tier performances. The weight of expectation, subtly different yet equally crushing.

Leah's performances reflected her growing resilience. Her rounds weren't flawless, certainly not the polished exhibitions of her earlier career. There were moments of hesitation, minor errors, and reminders of her vulnerability. But instead of crumbling under pressure, she learned to trust her instincts more and rely less on striving for perfection. This realization was pivotal, freeing her from the constant self-criticism and transforming her competitive mindset. She now focused on the process rather than the outcome, a subtle shift that yielded profound results. She focused on the rhythm, the fluidity of the jumps, and the connection with Whisper.

Yes, she made mistakes, but didn't let them define her. She learned from them, adjusting her strategy, refining her technique, and strengthening her bond with Whisper. Each minor victory, each clean jump, strengthened her confidence, building a resilience that extended far beyond the arena. She was learning not to compete, but to cope with the relentless pressure of the competitive circuit. This newfound ability, nurtured during the challenging time, enabled her to approach each round with newfound clarity, calmness, and focus, ultimately transforming her overall performance.

In the later rounds, it wasn't physical prowess that determined the winners, but their ability to handle the pressure and maintain their composure under intense circumstances. With each completed round, the pressure mounted on all competitors, and each competitor's powerful performance increased the tension, creating a thrilling competition. A palpable sense of high stakes hung heavy in the arena's air, the atmosphere crackling with the anticipation of the crowd, a feeling that all those in attendance could feel. The weight of expectation settled heavily upon Leah, a suffocating pressure that made it difficult for her to breathe and think clearly. Instead of succumbing to fear and becoming paralyzed, she had learned to redirect and harness that fear, transforming it into the fuel for her unwavering determination. Despite the immense pressure she was facing, she was resolutely determined to stay on course and not allow it to hinder her progress. She had journeyed a significant distance, traveling extensively and encountering many challenges along the way.

During one particularly challenging round, a tricky triple combination appeared to test Leah's horsemanship, and Whisper seemed to hesitate. It was a fleeting moment, a fraction of a second, but it was enough to unsettle her. She saw the doubt flicker in Whisper's eyes. But instead of succumbing to her anxieties, she instinctively lowered her center of gravity, adjusted her position, and coaxed Whisper over the jumps with a calm voice and a gentle touch. They were seamless, showing her newfound strength and confidence, leaving the audience spellbound by her grace and skill. This round showed her growth in pressure handling, demonstrating the impact of her resilience-building training regime.

The final round arrived, and the tension was almost unbearable. Leah was against Emma, of course, a rivalry that, despite the earlier understanding, had not entirely faded. Emma, looking composed but intense, was a picture of controlled power. The silent challenge, the unspoken acknowledgment of shared ambition, hung between them, heavy. This added to the excitement and intensity of the round. Leah could feel it, the undercurrent of competitiveness, a silent acknowledgment of their shared ambition and mutual respect.

Leah radiated a quiet calm. The earlier anxieties and paralyzing self-doubt had vanished, replaced by a calm certainty. She had found a profound calmness within herself, which was even more astonishing than her polished riding skills. This calm wasn't confidence, but a profound sense of inner peace — a self-assurance that extended beyond the competition. Her movements were fluid, almost effortless. She and Whisper were in sync, a unified team working in perfect harmony. They flowed around the course, each jump taken with precision and grace, leaving the crowd breathless. She didn't ride; she danced with Whisper, and her every movement perfectly matched her horse. This mutual trust, nurtured through hard work and dedication, led to a performance that transcended mere competition. It displayed a harmonious partnership, a testament to her resilience and growth as an equestrian.

The result was a tie—a perfect score for both riders. The judges conferred. A photo finish couldn't determine a victor. The atmosphere was electric with anticipation. The crowd held its breath. It was a rare and intense moment, a culmination of all the training and dedication, leaving both competitors visibly exhausted but filled with immense satisfaction. The event was a testament to their shared skills, showcasing their ability to overcome challenges and maintain exceptional control under tremendous pressure. Their unwavering focus, commitment, and resilience shone through during the competition.

Ultimately, they decided, but it almost didn't matter. Leah had already won, not the competition, but something far greater. She had overcome self-doubt, conquered fear, and embraced the power of resilience. She had discovered a strength within herself, forged in the crucible of adversity. Her transformation wasn't physical; it was a journey of self-discovery and emotional growth.

The tie, in its way, was a fitting conclusion. It celebrated not individual talent, but the extraordinary bond between horse and rider, and the unwavering support that had allowed her to flourish. It was a testament to her resilience, steadfast determination, and profound connection with Whisper. The journey, the setbacks, and the triumphs all contributed to the extraordinary young equestrian she had become. And as she stood there, hand resting on Whisper's smooth coat, Leah knew this was only the beginning. The competition had tested and challenged her, but ultimately empowered her to grow stronger, both as a rider and a person. The experience underscored the deeper meaning behind the sport and the resilience it demanded.

# Chapter 21: A Critical Decision

The celebratory atmosphere following the tie hung heavy, a shimmering veil over the lingering tension. The judges' decision, awarding the championship to Emma based on a technicality regarding a slightly quicker time in the final round, felt almost insignificant compared to the monumental shift within Leah. She had ridden flawlessly, pushing herself and Whisper to the absolute limits of their combined abilities. The judges' decision, however arbitrary, did not diminish the feeling of triumph that washed over her. It was a victory of self-discovery, a testament to the courage she had found within herself.

Later, alone in the stables' quiet solitude, the competition's afterglow slowly faded, replaced by a calm contemplation. Whisper nuzzled her hand, his warm breath comforting against her skin. The near-disaster with Whisper's leg, the initial panic, the grueling recovery, and the intense pressure of the competition coalesced into a profound sense of accomplishment. It wasn't about the ribbons or the accolades; it was about the journey, the transformation she had undergone.

The prestigious national championships were weeks away—a competition of unparalleled scale and intensity. It represented the pinnacle of youth show jumping, offering a chance to compete against the best riders in the country and a platform to showcase the talent and skills honed over years of dedication and perseverance. However, taking part meant subjecting Whisper to strain further, potentially jeopardizing his already fragile recovery.

The weight of this decision pressed heavily on Leah's shoulders. She paced the stable, the scent of hay and horse sweat filling her senses. Should she risk pushing Whisper to take part, potentially harming him in the process? Or should she withdraw, prioritizing his health and wellbeing over her ambition?

The choice felt agonizing, a battle between her aspirations and her responsibility for her beloved horse.

Mr. Henderson, her wise trainer, offered no simple answers. He watched her silently, understanding the conflict raging within her. He knew the importance of this decision; it spoke volumes about her growth, priorities, and the depth of her relationship with Whisper. He saw the internal struggle reflected in her eyes, a mixture of ambition and concern, a testament to the maturity she had gained.

"Leah," he finally said, his voice soft yet firm, "this isn't about winning or losing. This is about making a choice that reflects your values, understanding of Whisper, and ability to make hard decisions under pressure." His simple yet profound words cut through the indecision swirling within her mind. He wasn't directing her, but guiding her through self-reflection.

Leah spent the next few days immersed in a whirlwind of consultations. She sought opinions from veterinarians, physiotherapists, and fellow riders, weighing the potential risks and benefits. Each conversation added layers of complexity, presenting different perspectives that contributed to the gravity of her decision. She meticulously reviewed Whisper's medical records, poring over every detail, analyzing every subtle change in his gait and behavior. She spoke to Sarah, her friend, who listened patiently, offering support and a much-needed dose of reality.

Sarah's perspective offered an impartial external viewpoint. She pointed out the immense pressure Leah was under to perform, the expectation, and the potential consequences of withdrawing from such a prestigious competition. Sarah reminded Leah of the journey she had endured, her resilience, and her newfound self-belief. She also highlighted the significance of this decision in shaping Leah's future as a rider.

Yet Sarah also acknowledged the importance of prioritizing Whisper's health. She emphasized that the horse's well-being was paramount and that taking part in the national championships should not come at the expense of his long-term physical and emotional well-being. This was a stark reminder of her fundamental commitment as a rider and a testament to her growth in understanding the profound relationship between horse and rider.

Leah oscillated between ambition and responsibility. The thrill of competing nationally was immense, a dream she had long cherished. She pictured herself poised and confident, guiding Whisper through the course, the crowd's roar forming a symphony around them. The image fueled her desire, but the possibility of jeopardizing Whisper's recovery loomed like a shadow, a constant reminder of the potential consequences that could arise.

She finally spent a quiet afternoon in the field with Whisper, talking to him and sharing her dilemma. She traced the lines of his elegant face, feeling the subtle tremors beneath his skin. His enormous eyes seemed to sense her distress, reflecting her conflicting emotions. It was in those quiet moments, with only the wind rustling through the tall grasses as company, that Leah found her answer. The image of Whisper, healthy and happy, galloped effortlessly in a field, outweighing the allure of victory. The decision, when it came, was surprisingly straightforward and effortless. It wasn't a compromise but a resolute choice, born from a deep understanding of her values and unwavering love for her horse. She would withdraw from the national championships.

The relief that followed was immense, a wave washing over her, alleviating the burden of indecision that had weighed heavily on her for days. It was a hard decision that required tremendous courage and self-awareness, but it was the right one. She had prioritized Whisper's well-being over her ambition, a testament to her growth and maturity.

The news of her withdrawal rippled through the equestrian community. Some questioned her decision, suggesting it was a missed opportunity and a blow to her chances of national recognition. But others understood, praising her for her responsible approach and commitment to Whisper's welfare. It cemented her reputation as a talented rider and someone with integrity and compassion.

Leah didn't regret her decision. She viewed it as a pivotal moment in her journey. It showed her emotional maturity, ability to balance ambition with responsibility, and the strength of her bond with Whisper. She knew that genuine success in equestrian sports went beyond mere winning; it encompassed a holistic understanding of the relationship between horse and rider, a respect for their limitations, and an unwavering commitment to their well-being.

This turning point in her journey reinforced her understanding of the sport. It was a testament to her commitment, resilience, and ability to make hard decisions even under intense pressure. The lessons learned would serve her well in future competitions, shaping her approach to the sport and providing a solid foundation for future successes. Leah might have missed the national championships, but she knew her victory was far more profound. It was a victory of self-discovery, a testament to her commitment to Whisper, and a powerful lesson in the true meaning of sportsmanship and ethical decision-making. This critical decision marked a turning point in her competitive career and a significant step forward in her personal growth. She was ready for the next challenge, stronger and wiser than she had been before.

# Chapter 22: Confronting Emma

The decision to withdraw from the national championships settled like a quiet snowfall, blanketing the turmoil within Leah with a surprising calm. She had chosen Whisper's well-being over her ambition, which felt both profoundly right and strangely liberating. Yet, the lingering shadow of the competition and the lingering sting of Emma's victory still gnawed at her. She needed to confront that, to understand the full extent of her feelings and the nature of her rivalry with Emma.

The equestrian center buzzed with activity. The final preparations for the national championships were in full swing. Leah found Emma amidst the flurry, effortlessly graceful, as she adjusted her saddle pad, a picture of serene confidence. The sight of Emma radiating self-assuredness ignited a spark of something akin to defiance within Leah. This wasn't about the competition anymore, but about confronting the tension between them since the regional championships. It was about understanding their rivalry, roots, and potential evolution.

As Leah neared Emma, a familiar nervousness fluttered in her. The confrontation was unplanned, her words unprepared, but the urgent moment drove her forward. Emma's composed expression, a facade Leah knew for concealed strength and reserve, remained unchanged when she looked up.

"Emma," Leah started, her voice slightly shaky, "Is it possible to chat?"

With a perfectly sculpted eyebrow raised, Emma subtly conveyed amusement, or possibly disdain. "I guess we could," she responded calmly, her voice revealing none of her true feelings. They found a quiet corner, away from the bustling activity of the stables. The silence between them was thick with unspoken words, years of subtle jabs and competitive tension, the weight of their shared ambition and rivalry. Leah took a deep breath, gathering her thoughts.

"About the regional championships," Leah began, her voice gaining strength, "I know the judge's decision was based on a technicality, but..." She paused, searching for the right words. "But I felt cheated."

Emma's lips curved into a slight smile, a hint of mockery that stung Leah. "Cheated? Leah, you're a fantastic rider. You nearly took me down."

"Nearly isn't good enough," Leah retorted, her voice rising slightly.

"I trained harder than ever, pushing myself and Whisper to our limits. And I still lost because of something so trivial." Her voice cracked with frustration and something else–a vulnerability she hadn't expected. She wasn't challenging Emma's victory; she was challenging the system's unfairness, the arbitrary nature of the judging process, and the relentless pressure she felt to meet certain expectations.

Emma gazed at the other person, her mockery fading, replaced by a brief flicker of recognition or perhaps understanding, a glimpse of an unfamiliar emotion. She spoke more gently, acknowledging that the pressure affected them all. Drawing on her own experiences, she spoke with a composure that briefly faltered, betraying the underlying tension and stress she had been suppressing.

Leah's anger slowly abated, replaced by a tentative empathy. She had always seen Emma as her rival, a formidable opponent who stood in her way. But seeing Emma's momentary vulnerability revealed a different side–a person striving to maintain control under the immense pressure of the sport. Leah was acutely aware of the pressure.

"I know," Leah said, her voice subdued. "It's like a weight on your chest, suffocating you."

"Exactly," Emma nodded. "And the media glorifies winning, but they

We rarely acknowledge the sacrifices, dedication, and risk involved. We're all striving for perfection, but perfection is elusive."

The nature of their conversation changed, taking an unexpected turn. In their discussion, they delved into the unrelenting pressures inherent in their sport, the pervasive scrutiny they faced, the many sacrifices they had made, and the considerable emotional toll that this demanding lifestyle had exacted upon them. In their conversation, they delved into discussions about their families, the pressures and expectations placed upon them, and their intense yearning to show their capabilities as both skilled riders and powerful individuals. Delving into the intricacies of their intense rivalry, they acknowledged not only the fiercely competitive spirit that drove their actions, but also a quiet, unspoken respect for each other's exceptional skill and unwavering dedication.

For the first time, Leah saw Emma not as a rival to be defeated, but as a fellow competitor, a fellow warrior battling the same demons. They shared stories about their horses, the challenges they faced, and the joys of connecting with these magnificent animals. They spoke of training, of the setbacks and triumphs, the moments of pure exhilaration, and the crushing weight of defeat. The conversation was a bridge, spanning the chasm of their rivalry.

As their conversation wound down, a subtle shift had occurred. The rivalry remained, undoubtedly, but a newfound understanding and a grudging respect tempered it. The pressure remained, but now they acknowledged their shared burden. They were still competitors, but they were also comrades, bound by their shared experiences and a deep love for the sport. The conversation had cleared the air, revealing a more complex and nuanced dynamic than either had realized.

Leah walked away, unexpectedly feeling a sense of release. She still felt the sting of the judges' decision, but it no longer felt like a betrayal. Instead, it was a lesson, a reminder of the complexities of competition and the importance of self-awareness. Her confrontation with Emma changed the dynamic of their relationship. It deepened her self-awareness, providing her with valuable insights into her enduring pressures and strengthening her resolve to face future challenges.

Though she didn't win the national championships, her personal growth was a tremendous accomplishment, marking a significant change. Leah felt confident and prepared to face future challenges, even though the path ahead was uncertain and potentially difficult. The power she now held, born of quiet strength, surpassed any prize ribbon's worth.

# Chapter 23: Unexpected Alliances

The following days were a blur of activity, a whirlwind of preparations for the upcoming regional show. Despite her disappointment at missing the nationals, Leah found herself surprisingly focused. The conversation with Emma had been a turning point, a revelation. It wasn't the shared pressure, but the shared passion — the unspoken understanding of the dedication and sacrifice required to excel in this demanding sport. This new perspective changed her focus, enabling her to consider the broader context of the equestrian community beyond the immediate competition.

She gravitated towards other riders, those she might have previously dismissed as mere competitors. There was Chloe, a quiet, determined rider with a chestnut mare named Ruby, whose elegant style always caught Leah's eye. Chloe had an almost Zen-like calm in the saddle, contrasting with Leah's often-frazzled energy. They found common ground in their meticulous horse-care routines, swapping tips and tricks during the quiet moments between classes. Chloe's perspective, grounded in patience and perseverance, offered a refreshing counterpoint to Leah's impulsive approach.

Then there was Ben, a charismatic young rider with a flamboyant style and a mischievous grin. Ben, known for his daring approach to challenging courses, initially seemed the opposite of Leah's cautious nature. However, beneath the showmanship, Leah discovered a deep understanding of equine psychology and an almost intuitive connection with his horse, a spirited gray named Storm. Ben's candid advice about managing Storm's exuberance surprisingly helped Leah address some of Whisper's more stubborn habits. He taught her the importance of adapting riding techniques to suit each horse's unique personality, rather than rigidly sticking to a prescribed method. These exchanges weren't about technical advice; they were about building a rapport and camaraderie that transcended the competitive arena.

The drive to excel remained strong, but it was now tempered by a profound appreciation for the shared dedication and challenges faced by all riders. One evening, while watching the sunset over the sprawling equestrian center, Leah spoke with Sarah, a seasoned rider and mentor to many younger competitors.

With her decades of experience, Sarah had witnessed countless triumphs and heartbreaks within the sport. She shared stories of her struggles and defeats and how she had learned to overcome setbacks and find her rhythm again. Sarah's candidness offered a powerful reminder that success in equestrian sports was not solely about winning, but the ongoing process of growth, self-discovery, and resilience. She pointed out that true sportsmanship wasn't about fair play but about empathy and recognizing the community's shared struggles.

Leah's conversations with these other riders led to a more nuanced understanding of their pressures. They discussed the grueling training schedules, the financial burdens, the constant pressure to perform at their best, and the emotional toll of competition. These conversations revealed a shared humanity, a common ground often obscured by intense rivalry and the focus on individual success. She found herself less consumed by her anxieties and more aware of the shared challenges faced by her fellow competitors.

One afternoon, during a practice jump session, Leah witnessed a younger rider, Emily, struggle with a complex triple combination. Emily, visibly upset and frustrated, was on the verge of tears. While remembering her struggles with self-doubt, Leah approached Emily, offering encouragement and practical advice. She shared her techniques for approaching challenging jumps, emphasizing the importance of maintaining a calm and confident demeanor while building trust with the horse. Emily, initially hesitant, responded positively to Leah's support and eventually completed the jump with growing confidence.

This simple support fostered a sense of unity within the group of riders. It showed the fair play that extended beyond individual achievement. The shared experience of supporting each other and offering encouragement in times of difficulty created a stronger sense of community.

The atmosphere at the regional championships differed considerably from the tension Leah had experienced before. While the competition was still fierce, a sense of camaraderie among the riders had developed that hadn't been present previously. Leah approached her rounds with a newfound confidence, not only in her abilities, but also in the supportive environment she had helped cultivate. She performed exceptionally well, displaying both skill and grace.

The results were almost secondary; the genuine connections formed with her fellow competitors were far more significant. Leah discovered that genuine success in equestrian sports wasn't about winning medals or ribbons, but about the bonds forged, the lessons learned, and the shared experiences within a supportive and encouraging community.

The regional championship concluded with Leah securing a respectable placement, but more importantly, with a profound sense of satisfaction. The unexpected alliances she had forged with her fellow competitors had shifted her perspective, not only on the sport itself, but on her self-perception. She had confronted her insecurities, developed resilience, and discovered a community of support that enriched her experience far beyond the boundaries of competition. The journey, despite its challenges, had been profoundly rewarding.

Leah spent the quiet evenings following the regional championships in quiet contemplation. The exhilaration of competition had faded, giving way to thoughtful reflection on her journey. She had achieved personal growth far exceeding any tangible prize she might have won. The unexpected alliances had taught her the importance of empathy, the strength of shared struggles, and the powerful sense of belonging that emerged from genuine connection with her fellow competitors.

Her rivalry with Emma remained, but they redefined it. It was no longer a battle of egos, but a shared pursuit of excellence, fueled by mutual respect and an understanding of their challenges. Emma, too, had shown a subtle shift in attitude, a quieter appreciation for Leah's skill and determination. Their interactions were no longer marked by tension but by a calm acknowledgment of their shared journey within the demanding world of professional equestrian sports.

Looking ahead to future competitions, Leah approached them with a more balanced perspective. The pressure remained, of course, but it was now tempered by her resilience, strong support network, and understanding of the profound human element that thrived within the highly competitive world of show jumping. The victories and defeats would still matter, but the importance of building genuine connections, supporting fellow competitors, and embracing the journey had taken root in her heart. Her journey in show jumping had redefined the meaning of success for her, transcending the boundaries of ribbons and trophies to reach a deeper level of self-discovery and fulfillment. The quiet strength she carried within was a more substantial reward than any championship title could ever be.

# Chapter 24: Overcoming Obstacles

The weeks leading to the national championships felt like a relentless climb. The pressure, once a distant hum, had intensified to a roaring crescendo. Every training session was a grueling test of physical and mental endurance, each jump a potential stumble on the path to victory. Whisper, usually a willing partner, seemed to sense Leah's anxiety, her usual effortless grace replaced by moments of hesitation and uncertainty. Leah battled the challenging courses and her internal demons. Self-doubt, once subdued, gnawed at her confidence, whispering insidious lies about her abilities.

One particularly challenging practice session saw her stumble during a crucial triple combination. Although her fall resulted in only a minor scrape on her arm, and was not serious physically, the emotional impact was considerably more significant. The profound disappointment mirrored the doubts that had been quietly brewing beneath the surface of things, growing steadily stronger until they finally erupted. The immense pressure to succeed, to meet the high expectations and lofty ambitions of her family, proved almost too much for her to bear.

That evening, seeking solace, she found herself at the stables, stroking Whisper's soft muzzle. The quiet intimacy of the moment offered a momentary respite from the overwhelming pressure. In those quiet moments, she rediscovered the deep connection she shared with her horse, a connection that transcended the competitive arena. Sensing her distress, Whisper nuzzled her hand, a silent reassurance that calmed her racing heart.

Sarah, her mentor, noticed Leah's subdued demeanor the following day. She didn't pry, but offered a simple yet profound piece of advice. "Leah," she said, her voice gentle but firm, "remember why you started. Remember the joy, the connection, the sheer exhilaration of riding. The pressure is real, but so is the joy. Don't let the fear overshadow the passion."

Sarah's words resonated deeply. Leah realized that her focus had shifted entirely to the outcome, to the pressure of winning, and she had overlooked the inherent joy of the sport. She had lost sight of the love for the dance between her and Whisper, the harmonious blend of skill and intuition that made show jumping so captivating.

To regain her equilibrium, Leah consciously shifted her focus. She obsessed less over the competition and more time, simply enjoying the connection with Whisper. Long, quiet rides replaced intense training sessions, filled with the soothing rhythm of hooves on the grass and the gentle sway of the wind. These moments of reconnection brought a renewed sense of peace and confidence. It was like reigniting a dormant flame, rekindling her initial passion.

Meanwhile, the rivalry with Emma intensified. Confident and unburdened by self-doubt, Emma seemed to thrive on the pressure. Her performances were flawless, and her approach to the course was bold and calculated. Yet, Leah noticed something else in Emma's eyes, a hint of vulnerability beneath the façade of unwavering confidence. This realization subtly altered their rivalry. It was less a personal battle and more a reflection of the shared pressures and aspirations of the sport.

The national championships arrived, a whirlwind of anticipation and nervous energy. The atmosphere crackled with excitement, a potent mix of exhilaration and apprehension. Leah, however, approached her rounds with a newfound perspective. The pressure remained, but it lost its influence. Her focus was not on winning but on demonstrating her best, showcasing the connection she shared with Whisper, and enjoying the thrill of the competition.

The initial round showed a significant change in viewpoints and approaches. Leah moved with a fluid and graceful elegance across the stage, her performance a breathtaking demonstration of both exceptional skill and remarkable poise. Whisper's response was flawless, each movement precisely synchronized with the others in a display of remarkable precision. With the final fence cleared, a tremendous roar erupted from the crowd, an overwhelming wave of exhilaration washing over Leah as she completed the race.

The following round, however, presented a considerably more demanding set of challenges than the previous one. Near the middle of the course, a challenging jump of significant risk presented itself as a substantial and potentially insurmountable obstacle. Although deceptively simple in its design, the fence presented a challenge that demanded both perfect timing and impeccable balance from those who would construct it. The trial was not a measure of one's skill, but a test to determine the strength of their nerves and how well they could handle the pressure.

This significant jump in progress marked yet another critical juncture, representing a substantial shift in the overall trajectory. At the performance's start, the memory of her earlier, less-than-stellar practice resurfaced, bringing a prickling sensation of self-doubt that threatened to completely undermine her carefully constructed confidence. A wave of intense fear washed over her, momentarily paralyzing her and causing Whisper to hesitate before proceeding. Remembering Sarah's encouraging words, she inhaled deeply, redirected her attention to the task at hand, and painstakingly renewed the harmonious bond she shared with her equine partner.

Leah and Whisper approached the fence once more. A hush fell over the crowd as everyone held their breath in anticipation. Leah and Whisper successfully completed the jump, achieving undeniable success. She felt an immense wave of relief wash over her, a feeling of exhilaration so intense it was almost overwhelming. Having battled through significant internal conflict and ultimately triumphed over her fear, she reignited the fervent passion that had first captivated her and drawn her to the exhilarating world of show jumping.

With newfound confidence and grace, the participant executed the remaining rounds flawlessly, demonstrating significant improvement in skill and composure. The results, while respectable, were secondary to the profound sense of accomplishment Leah felt. She had not only navigated the intricate challenges of the national championships but also embarked on a journey of self-discovery, transforming her riding skills and her entire perspective on the sport.

The last event brought an unexpected twist. Emma, during her final round, experienced a sudden mishap. Her horse stumbled at a critical jump, resulting in a fall. The crowd gasped, a collective wave of concern rippling through the stands. Witnessing the incident, Leah felt a surge of empathy, forgetting the rivalry that had defined their relationship until that moment.

Emma, shaken but unharmed, struggled to compose herself. Leah immediately approached her, offering support and words of encouragement. This simple act of compassion was more significant than any victory or defeat. It marked a shift in their relationship, replacing rivalry with respect and understanding. Visibly moved by Leah's kindness, Emma accepted her support, and their shared experience created a bond stronger than any competitive tension.

The national championships ended not with a dramatic victory or a devastating loss but with a profound sense of shared experience. Leah's journey had been fraught with obstacles, both internal and external. She had faced self-doubt, endured setbacks, and navigated the fierce competition of the equestrian world. Yet, she had emerged stronger, more resilient, and with a newfound appreciation for the profound human element within the sport. The victories and defeats were now simply milestones on a journey of self-discovery, fueled by passion, resilience, and the supportive connections formed along the way.

# Chapter 25: Lessons Learned

The quiet hum of the stables, the scent of hay and leather, was a stark contrast to the cacophony of the championship arena. Days after the competition ended, the adrenaline faded, leaving behind a quiet introspection. The ribbons, the applause, the fleeting moments of glory were all fading memories now, replaced by a deeper understanding of herself and her journey. Sitting on the hay bale, her arm draped over Whisper's warm flank, Leah allowed herself to absorb the weight she'd experienced fully.

It wasn't the jumps, the flawlessly executed turns, or the breathtaking leaps over towering obstacles. It was far more profound than that. The national championships had been a crucible, a fiery test that had refined her as a rider and person. The lessons learned found their place not in scorecards or trophies, but in the fabric of her being.

She realized the importance of self-compassion was a significant lesson. She had been so focused on external validation, on the approval of her family, her coach, and even her rivals, that she had lost sight of her inner voice. The pressure to succeed, to live up to the expectations placed upon her, had almost crushed her. The fall during practice and the moment of doubt at the crucial jump were not failures, but opportunities for self-understanding. They were reminders that perfection is an illusion, and that it's okay to stumble, fall, and rise again.

This self-compassion extended to her relationship with Whisper. Before the championships, she focused entirely on performance, achieving zero faults and winning the coveted blue ribbon. She had almost forgotten the sheer joy of the partnership, the unspoken communication, the intuitive connection between them. The quiet rides, gentle grooming sessions, and moments spent together had been crucial in rekindling that connection. It was as if she had finally heard Whisper's voice clearly, silencing the noise of her anxieties.

Another vital lesson was the power of perspective. The national championships weren't about winning or losing; they were about the journey, about the challenges overcome, about the growth experienced. The competitive fire still burned within her, but a deeper understanding of the human element of the sport now tempered it. A shared experience had transformed the rivalry with Emma, initially characterized by tension and animosity,. Emma's fall had stripped away the artificial barriers that had separated them, revealing a shared vulnerability and a mutual respect. The act of supporting Emma, offering a hand in her moment of distress, had been more rewarding than any victory could have been.

This realization highlighted the interconnectedness of the equestrian world. It wasn't simply a collection of individuals vying for individual glory; it was a community of riders, trainers, horses, and families bound by a shared passion for the sport. The support she'd received from Sarah, her unwavering belief, and her gentle guidance had been instrumental in her success. The Camaraderie among competitors and a mutual understanding of the pressures and challenges created a sense of belonging that extended beyond the competition itself.

Leah also learned the crucial importance of balance. The intense focus on training and the relentless pursuit of perfection had led to burnout and self-doubt. She understood that balancing physical demands with mental well-being and personal fulfillment was crucial for her and Whisper's success. The long, quiet rides, the mindful moments spent tending to Whisper, and her renewed focus on her mental health were not distractions, but essential components of her training regimen.

Her relationship with her family had also undergone a subtle shift. The pressure to uphold the family's legacy had been immense, but now she understood their expectations stemmed from love and pride. Her achievements did not validate her worth in their eyes; they reflected her dedication and passion. She could now approach her riding with personal fulfillment, rather than trying to fulfill external expectations.

The national championships, with all their difficulties, had been more than a competition; they had been a transformative experience. It wasn't about the final standings on the leaderboard, but the personal growth that had unfolded during her journey. She faced her fears, acknowledged her vulnerabilities, and became stronger and more self-aware. She learned that ribbons and trophies don't always measure the most significant victories; lessons learned, challenges overcome, and connections forged do.

The memory of the last jump, which had tested her courage and skill, now brought a smile to her face. It wasn't a technical feat, but a symbol of her transformation. The hesitation, doubt, and ultimate triumph were all integral to her narrative of growth. She had learned that it wasn't about eliminating fear but facing it, acknowledging it, and moving forward with courage and determination.

The pressure hadn't vanished entirely, but it felt different now. It was no longer a crushing weight, but a challenge to be embraced, a source of motivation rather than anxiety. The joy of riding, the deep connection with Whisper, and the shared passion with other riders were the aspects that now fueled her pursuit of excellence. Her victories were now more profound, rooted in self-acceptance, perseverance, and an unwavering belief in herself and her capabilities.

The lessons learned at the national championships extended far beyond the confines of the show jumping arena. They were lessons about resilience, self-compassion, and the power of human connection. They were lessons about balancing ambition with self-care, embracing both victory and defeat, and finding joy in the journey itself. These lessons would stay with her long after the final ribbon was awarded, shaping her as a rider and a person, ensuring she would meet future challenges not only with skill and determination, but also with the unwavering self-belief she had so meticulously forged in the crucible of competition.

The setting sun cast long shadows across the stables, painting the scene in hues of orange and gold. Resting gently on Whisper's flank, Leah felt a profound sense of peace. The championships were over, but her journey had begun. She carried within her the memory of the competition and the wisdom gained from it, the self-awareness nurtured, and the strengthened resolve to continue her journey with grace, courage, and an unyielding love for the sport she so passionately embraced. The quiet strength she felt within, the subtle confidence that bloomed from overcoming her internal struggles, proved to be a more satisfying victory than any championship trophy ever could be.

# Chapter 26: Preparation for the Finals

A crisp morning air, biting with a chill that penetrated Leah's very bones, stood in stark contrast to the fiery intensity that burned within her. With the final rounds drawing near, a feeling of daunting challenge mixed with exhilarating anticipation filled the air. Months of dedicated training culminated in the national championships, a competition that helped her hone her riding skills and become a more accomplished equestrian. In the previous rounds, a flurry of activity unfolded, a whirlwind of adrenaline-pumping moments, a seamless blend of perfectly executed turns and daring leaps that showcased incredible athleticism. As the ultimate prize came into view, a new and heavier pressure bore down—the immense pressure of expectation, the heavy weight of their legacy, and the crushing fear of failure.

She had learned, however, that fear wasn't the enemy. Fear was a companion, a shadow she could acknowledge, understand, and ultimately transcend. The fall at the center fence in the earlier rounds, the crushing disappointment, had been a pivotal moment. It hadn't broken her; it had tempered her, revealing the resilience she hadn't known she possessed. She'd spent the intervening days not dwelling on the past, but dissecting it, understanding where she faltered and how she could improve. This wasn't simply about technical skill; it was about harnessing the power of her mind, about controlling the narrative within.

This realization had prompted a radical shift in her preparation strategy. She abandoned the relentless, almost obsessive focus on physical training that had previously led to burnout. Instead, she incorporated mindfulness techniques, spending quiet moments meditating by the stables, focusing on her breath, and clearing her mind of the racing thoughts and self-doubt that threatened to overwhelm her. Practicing visualization, she pictured herself flawlessly navigating the course, each jump a seamless transition and each turn a precise maneuver. She wasn't merely visualizing the physical aspects; she imagined the feeling of absolute control, the unwavering confidence, the calm strength that emanated from within.

Sarah, her ever-supportive coach, had been instrumental in this mental recalibration. Sarah understood that success in showjumping wasn't solely dependent on physical prowess; it was a symphony of mind and body, a delicate dance between skill and mental fortitude. Together, they meticulously reviewed the course map, analyzing each jump and turn, identifying potential pitfalls, and developing strategies to overcome them.

Sarah worked with Leah on her mental game, reinforcing the importance of positive self-talk, focusing on what she could control, and letting go of what she couldn't. They discussed managing pressure, dealing with distractions, and maintaining focus amidst the noise and excitement of the competition. This holistic approach treated Leah's mental preparation as an integral part of her physical training.

Leah's relationship with Whisper had also undergone a subtle yet significant transformation. She had shifted from viewing Whisper solely as an instrument for achieving zero faults—a tool for winning—to viewing her as a partner, a confidante, and a friend. She spent hours grooming Whisper, softly stroking her coat, whispering words of encouragement and reassurance. The quiet moments together were more than physical care; they were acts of nurturing, fostering a deeper level of communication and trust. Their connection extended beyond the arena; a bond transcended the competitive pressures, becoming a source of strength and solace.

The mental preparation involved more than visualization and mindfulness; it included a thorough analysis into strategy. Leah analyzed her previous rounds, pinpointing her weaknesses and capitalizing on her strengths. She had identified a tendency to overthink, to allow doubt to creep in at crucial moments. Now, she was working on actively silencing that inner critic, replacing negative self-talk with affirmations and self-encouragement—her strategy involved focusing on the process, not the outcome. The goal wasn't simply to win; it was to execute each jump with precision and grace, to maintain her focus and composure, to ride with unwavering confidence, and to ride as one with Whisper.

The other competitors, once perceived as rivals, now felt like comrades in arms, sharing a common battlefield and enduring shared challenges. Even Emma, with whom she had had a fraught relationship, seemed less like an adversary and more like a fellow competitor striving for excellence. The shared experience of intense pressure and relentless challenge fostered grudging respect, a recognition of their shared dedication and passion. The sight of Emma's unwavering dedication during her practice sessions was a source of inspiration.

Leah observed other riders, studying their techniques, and learning from their successes and failures. The arena was a vibrant hub of activity, a tapestry of skill, determination, and unwavering passion. She noted how some riders approached their mounts with an almost reverent calmness, a quiet confidence that radiated from them. Others were fiery and energetic, their energy seemingly infectious. She noted the subtle differences in riding styles, the variations in approaches, and the distinct ways riders managed their mounts. This wasn't merely competitive observation; it was a learning opportunity to glean wisdom from experienced riders.

As the final rounds approached, frenetic activity and peaceful introspection filled the intervening days, a kind of symphony of action and quiet reflection. Within the stables, a low, almost tangible energy thrummed, a palpable blend of eager anticipation and the nervous excitement that hung heavy in the air. In the air, a potent mixture of the equestrian world was present; the freshly cut hay, the aroma of leather, and the scent of liniment all blended together. A vibrant energy filled the air, simultaneously exhilarating and nerve-wracking, creating a potent atmosphere that was both electrifying and unsettling. Leah discovered that a harmonious blend of intense physical training and periods of tranquil meditation was essential for her well-being, a discovery that underscored the significance she now placed on mental health.

The final preparations included a meticulous check of Whisper's equipment, a grooming session, and a final walk-through of the course. She mentally rehearsed the course, visualizing each jump and turn, planning her strategy, and preparing for any challenge. She ran through the process in her head, not the jumps, but the dismounts, the turns, the hand signals, the communication with her horse. This wasn't a mere technical rehearsal, but a form of mental meditation, a method of grounding herself in the present moment and connecting deeply with her partner.

The night before the finals was tense, but Leah found peace in a quiet evening with Whisper. She brushed Whisper's coat until it gleamed, whispering reassuring words and feeling the steady rhythm of Whisper's breathing against her palm. The bond was deeper now, forged in the fires of competition and strengthened by shared struggles. The sense of partnership was palpable, a shared understanding that transcended words. They were a team, ready to face whatever lay ahead. She knew that whatever happened in the arena the next day, their journey was greater than any result. This was a testament to their resilience and bond, built on trust, mutual respect, and unwavering support. In these quiet moments of connection, they built true strength, a strength that would serve them well in the challenges ahead. And as she drifted to sleep, she felt a calm confidence, a settled certainty that had been absent weeks earlier. She was ready.

# Chapter 27: Increased Stakes

The air crackled with anticipation as Leah and Whisper entered the warm-up arena. The familiar scent of sawdust, sweat, and horse liniment hung heavy in the air, a heady mix that usually spurred her adrenaline, but today, it felt different. A calm, focused energy pulsed through her, a quiet confidence that had taken root over the past few days. She lacked physical preparedness.

The warm-up wasn't about pushing Whisper to her limits; it was about establishing a rhythm, a connection, a silent conversation between horse and rider. Each movement was intentional, and each jump was a measured arc. Whisper responded with an effortless grace, her movements fluid and powerful, a testament to the trust and understanding that had blossomed between them. Leah felt the subtle shifts in Whisper's body, anticipating her every move as they danced into a seamless partnership. They were a single entity, two hearts beating as one.

The final round was a different beast altogether. The earlier rounds had been a test of skill, a measure of technique. This test of character was a crucible that would forge the staunch champions. The course was more demanding, with jumps higher, turns tighter, and obstacles more intricate. The crowd's roar, usually a source of distraction, was now a muted hum, a distant backdrop to the intense focus within her. Leah had learned to tune out the external noise, to focus solely on her connection with Whisper.

When they stepped into the arena, the atmosphere crackled with a tension so thick you could cut it with a knife, a silent hum of anticipation hanging in the air before the contest began. The intense gaze of the judges, the barely audible whispers from the spectators, and the expecting stares of her fellow competitors—these external pressures, once so significant, now faded into insignificance. Completely enveloped in a sphere of intense concentration, she devoted her attention solely to the rhythmic beat of Whisper's hooves against the ground, the sensation of the reins in her hands, and the unbreakable bond they shared.

The first few jumps were a blur. They moved as one, a perfectly coordinated team, each jump a harmonious blend of skill and trust. Leah noticed Emma in the stands, her usual arrogant smirk replaced with a look of quiet respect. This was a battle not against the course, but against self-doubt, against the weight of expectations, against the specter of failure.

But the course was unforgiving, designed to test even the most seasoned riders. Many found the center fence a fearsome obstacle, but she negotiated it effortlessly. Leah's breath was steady, her heartbeat even, her mind clear and focused. Winning wasn't on her mind; The triple combination of fences loomed ahead–a treacherous sequence of jumps that demanded exceptional precision and timing. A single misstep could be catastrophic, throwing off their entire rhythm and potentially costing them the competition.

Leah felt a flicker of apprehension, a familiar shadow of doubt creeping into her mind. But she quickly banished it, replacing it with positive self-talk. whispered almost inaudibly over the pounding rhythm of the horses' hooves, she conveyed the sheer number of rehearsals, said, "We've rehearsed this countless times, more than I can remember. "We're committed to completing this task."

And, following through on their intentions, they indeed completed the task. They sailed through the triple combination with effortless grace, each jump seamless and precise. The crowd erupted in a roar of applause, but Leah remained focused, her attention fixed on the last obstacle—a towering obstacle, the ultimate test of courage and skill. This was it—the culmination of months of rigorous training, the moment of truth. The pressure was immense, yet Leah felt strangely calm. She was no longer consumed by fear; a powerful surge of determination fueled her. This wasn't about winning; it was about proving herself, not to anyone else, but to herself.

As they approached the obstacle, Leah took a deep breath, visualizing the jump in her mind and feeling the rhythm of Whisper's strides, anticipating every movement. They soared over the fence, a magnificent leap that defied gravity, a testament to their skill and unwavering bond. They landed gracefully, their movements smooth and controlled, a perfect ending to a near-perfect performance.

The silence that followed was deafening, punctuated only by the rhythmic pounding of their hearts. Leah looked at Whisper, her eyes filled with pride and a sense of relief. They had done it. They had faced the ultimate test and emerged victorious. The moment of triumph was bittersweet, marking the culmination of years of dedication and bringing to a close a long, challenging, and emotional journey. It was more than a win; it was a symbol of her personal growth, a testament to her resilience, and a celebration of their unbreakable partnership.

Officials announced the results shortly after, and the tense wait amplified the emotions in the stadium. The judges painstakingly tallied the scores, each point a testament to the meticulous precision of the performance. Then, the announcer's voice boomed across the stadium, clear and decisive: "And the National Show Jumping Championship winner is...Leah and Whisper!"

The eruption of applause was deafening, a tidal wave of cheers that washed over Leah. She felt a surge of emotion, a potent cocktail of joy, relief, and overwhelming pride. Tears welled in her eyes, not tears of self-pity or frustration, but of pure joy. This wasn't a win; it was a testament to her dedication, resilience, and unwavering bond with Whisper.

Emma approached her, her usual aloofness replaced by a genuine smile. "Congratulations, Leah," she said, extending her hand. "You rode brilliantly."

Leah shook her hand, a wave of camaraderie washing over her. The rivalry, the tension, the pressure—all of it had fallen away, replaced by a shared understanding of the dedication, the passion, and the sheer hard work that had led them both to this moment. They were competitors, yes, but they were also fellow riders, united by their love of the sport and their mutual respect for the art of equestrianism.

The victory celebration was a whirlwind of congratulations, handshakes, and well-wishes. But Leah found herself drawn back to Whisper, her loyal companion, her partner, on this extraordinary journey. She spent the evening grooming Whisper, whispering words of gratitude and affection, cherishing the moment, the victory, and the unbreakable bond they shared. The win symbolized their shared journey, a testament to their unwavering trust and mutual respect, a celebration of their shared passion, and a promise of many more victories to come. It was a moment that cemented their bond, strengthened their partnership, and solidified their place in the annals of equestrian history.

# Chapter 28: Facing the Pressure

The final rounds loomed, a stark contrast to the relatively relaxed atmosphere of the earlier heats. The air thrummed with a unique energy now—a palpable tension that vibrated through the ground, Whisper's legs, and into Leah's body. It wasn't the nervous energy of the earlier rounds; this was a more profound, more visceral feeling, a pressure that pressed down on her chest, a weight she could almost feel physically. The earlier rounds had been a warm-up, a gentle introduction to the intensity of the competition. This was the main event, the culmination of months, years even, of relentless training and unwavering dedication. Here, champions were made or broken.

The course itself seemed to have shifted, transformed by the heightened pressure. The jumps, previous obstacles to be navigated, now loomed like formidable adversaries, their heights amplified, their distances stretched. Each fence held a silent challenge, whispering doubts into Leah's ear, tempting her to falter and yield to the immense pressure on her. She could almost hear the crowd's whispers, a low hum of anticipation that edged toward a threatening murmur. The usual supportive cheers now felt like a judgment, each sound magnifying the stakes, the weight of expectation.

Leah closed her eyes momentarily, taking a deep, steadying breath.

The scent of Whisper's warm breath and the feel of her soft muzzle against her hand anchored her back to the present, grounding her in the shared reality of their partnership. This wasn't about her; it was about Whisper, too. Whisper, who had trained alongside her, endured the long hours, the sweat, the frustration, and the unwavering dedication. Whisper deserved her best, and Leah would give nothing less.

Opening her eyes, Leah focused on her breathing, a technique she'd honed over years of practice. Inhale, exhale. Each breath was a conscious act, a deliberate decision to maintain her composure, to push back against the tidal wave of pressure threatening to overwhelm her. She visualized the course, each jump a clear image in her mind's eye. She replayed each jump, turn, and movement in her mental theater, reinforcing her muscle memory and confirming the choreography of their performance. It wasn't about the physical execution, but the mental preparation, the unwavering focus, and the silent conversation between horse and rider.

The first jump approached, a deceptively simple vertical that many riders would easily clear. Yet, for Leah, it held symbolic weight, representing the entire competition and serving as a gateway to success or failure. She didn't rush; she felt the measured rhythm of Whisper's strides, the subtle shift in her weight, the anticipation of the jump. It wasn't a contest of speed, but a dance of precision and trust. They took off, a perfect arc against the backdrop of the stadium's hushed anticipation, the silence broken only by the soft thud of their hooves and the whisper of the wind.

One jump flowed seamlessly into the next. Each fence tested skill and composure, yet Leah met them all with unwavering focus and determination. She was no longer aware of the crowd, of the judges, of her competitors. Her world had shrunk to the small space between her and Whisper, a shared world of rhythm, trust, and unwavering dedication. Once a daunting obstacle, the course was now a series of challenges to be met, each a testament to their partnership, training, and mutual respect.

The next few jumps were formidable, and yet they loomed and often tripped up even the most experienced riders. The glistening surface of the jumps reflected the stadium lights, creating an almost surreal atmosphere and enhancing the challenge. But Leah felt no fear; there was only a quiet determination and a calm confidence in her abilities, as well as Whisper's unwavering strength. They approached the jump, a steady rhythm guiding their movement, and sailed over it with effortless grace, the obstacle barely disturbing their momentum.

The triple combination was a challenging sequence of jumps requiring perfect timing, precision, and coordination. A misstep here could cost them the entire competition. Leah felt the familiar flicker of apprehension, the shadow of self-doubt whispering in her ear. But she pushed it aside, reinforcing her mental strength with positive self-talk. "We've done this a thousand times," she murmured, the words a silent affirmation of her confidence and their shared mastery.

They cleared the triple combination effortlessly, confidently blending techniques in each jump. The crowd erupted, their cheers a wave of sound that washed over Leah, but she remained focused, her mind locked onto the last obstacle—the towering obstacle, the ultimate test. This was it—the culmination of all their efforts, the moment of truth, the last hurdle before victory or defeat.

The obstacle seemed to grow larger as they approached, and the pressure was almost palpable. Leah could feel the weight of expectation, the crowd's pressure, and the intensity of the competition. Yet, amidst the overwhelming pressure, a calm determination settled over her. It wasn't about winning or losing anymore; it was about performing to the best of their abilities, honoring their training, partnership, and mutual dedication.

They approached the last jump, a silent communion between horse and rider. There was no hesitation, no doubt, only a shared understanding of their goal, a mutual trust that transcended words. They took off, a majestic leap that seemed to defy gravity, a testament to their unwavering skill and bond of power. They landed gracefully, their movements fluid and smooth, their rhythm unbroken, their shared victory secured.

The silence that followed was deafening, the anticipation hanging heavy in the air. The judges tallied the scores, each reflecting their skill and precision. Then, the announcer's voice boomed across the stadium, the words echoing the culmination of years of dedication, sweat, and perseverance: "And the winner of the National Show Jumping Championship is... Leah and Whisper!"

The crowd's roar was deafening, a wave of sound that washed over Leah, carrying her along in a tide of joy and relief. It was more than a win; it was a validation of their journey, a symbol of their unwavering commitment, and a testament to the unbreakable bond between horse and rider.

# Chapter 29: Unexpected Twists

The euphoria of that perfect last jump quickly dissipated as the judges began their deliberation. The silence stretched, thick, each second feeling like an eternity. Leah, still perched on Whisper, felt a tremor of unease. The seemingly flawless run, the effortless grace, the perfect landing—all of it felt fragile, vulnerable, under the scrutinizing gaze of the judges. The crowd, previously a roaring tide of cheers, had retreated into a hushed expectancy, a palpable tension hanging in the air thick enough to cut with a knife. This wasn't the celebratory silence of victory; this was the tense anticipation of judgment.

Then came the announcement, but not the one she expected. Instead of the triumphant declaration of victory, the announcer's usually confident voice held a hesitant quality. "There's been... a slight complication," he said, his voice laced with an unusual uncertainty. "One of the judges... well, let's say there's a discrepancy in the scoring."

A murmur rippled through the crowd, a wave of confusion and speculation washing over Leah. A discrepancy? What discrepancy? The sweet taste of victory had turned sour. The judges huddled together, their hushed conversation a source of growing anxiety. Leah felt an icy dread creep into her heart, a chilling premonition of what might come next.

The tension in the air was unbearable. Sensing her rider's anxiety, Whisper shifted restlessly, her breath hot against Leah's cheek. The minutes stretched into an eternity, each second feeling like a hammer blow against Leah's already frayed nerves. She tried to remain calm, to project an air of composure, but her heart hammered against her ribs, a frantic drumbeat against the suffocating silence.

Finally, the head judge stepped forward, his expression grave. He announced that a re-judge was necessary because of a disagreement over the scoring of the last obstacle. A re-judge! The word itself felt like a slap in the face, a cruel twist of fate threatening to unravel everything for which she had worked. Her flawless run and perfect landing are now reduced to a potential source of contention, a subject of debate.

The news spread like wildfire through the crowd, igniting speculation and murmurs. The previously supportive crowd was now divided, a sea of opinions and conflicting interpretations. Some whispered about bias, while others discussed a flawed judging system, and some even dared to suggest foul play. The storm's eye held Leah, making her the target of sympathy and suspicion.

The re-judge was even more nerve-wracking than the initial judging. Leah and Whisper were called back into the arena, the atmosphere now heavy with uncertainty and speculation. The pressure was immense, a suffocating weight that threatened to crush her. This wasn't about winning anymore; it was about proving her worth, validating her skills, and silencing the doubts creeping into her mind.

The re-judge wasn't a repeat of the previous run, but a different beast entirely. The judges were more observant and scrutinizing, their gaze piercing, their attention focused on every detail, every movement, and every nuance of the horse and rider partnership. It felt more like an interrogation than a competition.

Leah fought to maintain her composure, to push down the rising tide of anxiety. She focused on Whisper, feeling the rhythmic beat of her heart, the steady rise and fall of her breath, seeking comfort and strength in their shared connection. Whisper, surprisingly, seemed unfazed, her usual calmness and steadiness reassuring Leah in her trepidation.

They cleared each jump with precision and grace, a testament to their unwavering training and partnership. But this time, it wasn't the effortless, almost magical performance of the first run. This was a performance fueled by determination, a fierce resolve to prove themselves, to dispel the doubts cast upon their victory.

When they landed from the last obstacle, the silence was even more pronounced than before. The judges' expressions were unreadable, their faces masks of impassivity, their silence a source of agonizing uncertainty. The wait felt eternal, the tension so thick it was almost suffocating.

Then, the verdict came, and it was even more unexpected than the initial delay. The head judge announced that there was no clear winner. The scores were too close to call—a tie. A collective gasp went through the audience. A tie? It was unprecedented, a situation that no one had foreseen, a twist that no one had expected. The competition rules stated that a tie-breaker round would be necessary.

Competitors faced an extremely demanding tie-breaker that tested their endurance, precision, and nerve to the absolute limit. The shorter course, with its intricate turns and high jumps, presented a more technically demanding challenge, specifically designed to separate the most skilled competitors from the rest. The challenge tested not only physical prowess, demanding both skill and stamina, but also the participants' mental resilience, pushing them to their limits.

Leah felt the pressure build, but this time, it was different. This wasn't the pressure of expectation, but the pressure of competition — the thrill of the chase. She found a new level of focus, a clarity that had previously eluded her. The uncertainty of the previous rounds had paradoxically sharpened her senses, making her more aware, alert, and determined.

Whisper responded to her energy, her movements fluid and powerful, her spirit as fierce and determined as Leah's own. They worked together as one, a seamless blend of horse and rider, and each jump is a testament to their partnership, training, and unwavering resolve.

They flawlessly completed the tie-breaker, with each jump executed perfectly and their movements showcasing remarkable skill and grace. The crowd erupted in cheers, a wave of sound that washed over Leah, a testament to their stunning performance.

The judges' scores were quick this time. There was no delay, no discrepancy, no uncertainty. The verdict was clear, decisive, and triumphant. The announcer's voice boomed across the stadium, carrying the weight of expectation and relief: "And the National Show Jumping Championship winner is... Leah and Whisper!"

This time, the cheers were louder, more jubilant, more heartfelt. It wasn't a victory, but a validation — a testament to their resilience, skill, and unbreakable bond. Many unexpected events and challenges only strengthened their resolve, forging an even stronger partnership. The hard twists and tie-breaker spurred Leah's development, highlighting her strength and resilience.

# Chapter 30: Showcasing Skill

The crowd's roar was a physical force, a wave crashing over Leah as she and Whisper entered the arena for the tie-breaker. The air crackled with anticipation, a palpable energy that was both exhilarating and intimidating. This instance was unlike any other round, setting it apart from the rest. This was the culmination of years of dedicated training, the fulfillment of a lifelong dream, and the defining moment of her life's preparation. However, this moment, while exciting, also carried a significant risk of immense and devastating letdown, a possibility that weighed heavily on everyone involved.

The course was a masterpiece of technical difficulty, a series of intricate turns and challenging jumps designed to test the limits of horse and rider. There were tight turns that demanded precise control, demanding a delicate balance between speed and accuracy. The fences were higher and more demanding, each requiring a perfect approach, takeoff, and landing. One mistake, one hesitation, could be catastrophic.

Leah felt a surge of adrenaline, a thrilling cocktail of fear and excitement coursing through her veins. Whisper, sensing her rider's energy, responded in kind. Her muscles were tense but relaxed, her breath steady, her ears pricked, and her gaze sharp and focused. They were a team, a single entity, their minds and bodies working in perfect harmony.

The starting gun fired, and a sharp crack cut through the expectant silence. Leah urged Whisper forward, her legs gripping the saddle, her body poised and balanced, her eyes scanning the course ahead, anticipating every turn, every jump. Whisper responded with breathtaking power and grace, her movements fluid and precise, her strides powerful and controlled.

The first few jumps were a blur of motion, a breathtaking dance of horse and rider. Leah felt an almost primal connection to Whisper, their movements symbiotic and intuitive. They executed each jump with perfect timing and precision, a testament to their unwavering training and deep bond. The crowd roared with its approval, a wave of sound that fueled Leah's determination, pushing her to push harder, ride faster, and strive for perfection.

The course challenged skill, stamina, and mental endurance. The pressure mounted with each passing jump, the tension growing, the stakes higher. Leah could feel the fatigue seeping into her muscles, the burn in her thighs, but she pushed on, fueled by her desire to win and prove herself.

The next obstacle was a triple bar, notoriously difficult even for seasoned riders. It required a perfect approach, precise timing, and the unwavering trust between horse and rider. Leah took a deep breath, focusing on the jump, concentrating on the rhythm, the flow, the subtle nuances of Whisper's movements.

They cleared the triple bar with a perfect leap, their movements flawless, their synchronization breathtaking. The crowd erupted in cheers, a tidal wave of sound that seemed to lift Leah and Whisper off their feet.

The last jump loomed ahead, a towering obstacle, an ultimate test of nerve and skill. It was a challenge that would decide everything. Leah felt a surge of anxiety, a tremor of doubt, but she quickly pushed them aside. She focused on Whisper, feeling the rhythm of her breath, the beat of her heart, the steady power of her muscles.

They approached the last obstacle, their rhythm perfect, their partnership unwavering. Leah adjusted her position, leaning slightly forward, a guiding Whisper with a gentle touch, her body working in perfect unison with her horse. They took off, soaring through the air in a graceful arc of power and precision.

They landed with a soft thud, Whisper's four legs hitting the ground with a perfect, balanced landing. Leah felt a surge of relief and exhilaration washing over her. They had done it. They had conquered the course, overcome the challenges, and achieved victory.

The silence that followed was deafening, a stark contrast to the previous roar of the crowd. The judges huddled together, their faces impassive, their expressions unreadable. Leah waited, her heart pounding, the tension in the arena thick enough to cut with a knife.

The wait felt endless, drawn out in the silence. Every muscle in her body was taut, each breath shallow. Whisper stood quietly beside her, sensing her rider's apprehension. Her stillness calmed the overwhelming tension.

Finally, the head judge stepped forward, his expression unreadable, his voice clear but calm. "And the winner," he announced, his voice carrying across the hushed arena, "of the National Show Jumping Championship is... Leah and Whisper!"

The roar that erupted was deafening. A wave of pure exhilaration engulfed Leah. She felt tears welling up in her eyes—tears of relief, tears of triumph, tears of pure joy. Whisper nudged her gently, her warm breath comforting against her cheek.

The victory wasn't about the championship title; it was about overcoming self-doubt, facing challenges head-on, and forging an unbreakable bond with her horse. It was a testament to years of dedication, hard work, perseverance, and the unwavering support of those who believed in her. This victory culminated everything she had worked for, a dream realized, a testament to the power of passion, dedication, and the extraordinary bond between horse and rider. It was a moment etched in time, a memory to be cherished forever.

The journey had been arduous, filled with setbacks and self-doubt. Still, it had ultimately led to this triumphant moment that validated every sacrifice, struggle, drop of sweat, and tear shed along the way. It was a victory not for Leah but for Whisper, their partnership, trust, and unbreakable bond. The applause thundered around her, a sea of faces blurring into a joyous blur, but all she saw was Whisper, her magnificent, faithful partner, the source of her strength, solace, and triumph.

# Chapter 31: The Decisive Moment

The last jump loomed, a stark, imposing obstacle that grew larger with each stride Whisper took. It wasn't a jump; it was a monument to the pressure, the anxiety, the sheer weight of expectation that had been building since the competition began. A silence so profound it was almost unbearable hung over the arena, punctuated only by the rhythmic thudding of Whisper's hooves as they softly affected the ground and the frantic, erratic beating of Leah's heart within her chest. An almost palpable tension hung heavy in the air, a silent pressure that vibrated with unspoken anxieties and simmering emotions.

Leah focused, her mind a laser beam trained on the obstacle ahead.

She could almost feel the eyes of the judges, the scrutiny of the crowd, the silent judgment of her rivals. Yet, paradoxically, she also felt a strange calm settle over her. The adrenaline had subsided, replaced by a focused intensity that sharpened her senses, heightened her awareness. It was as if the world had narrowed, all extraneous thoughts and distractions fading away, leaving only her, Whisper, and the jump.

Sensing the shift in Leah's demeanor, Whisper responded with her quiet strength. Her breath was slow and deep, her muscles poised and ready, her attention laser-focused on the task at hand. Their connection, already profound, deepened in this moment of shared intensity. It was more than a partnership; it was a communion, a silent understanding transcending words and emotions. They were one entity poised on the precipice of victory or defeat.

The approach was crucial, a delicate dance of balance and precision. If they moved too quickly, they might lose control and fall. Too slow, and they'd lose momentum, leaving them vulnerable to a hesitant takeoff and a clumsy landing. Leah guided Whisper with imperceptible shifts in her weight, legs, and hands and a symphony of subtle cues that only her horse could understand and respond to. Leah measured each stride, made each movement deliberate, and synchronized each breath.

The crowd held its breath. The anticipation was palpable, thick, and heavy, hanging like a tangible substance in the air. Even the wind seemed to hush, as if respecting the sacredness of this decisive moment. Leah could feel the weight of every eye upon her, the silent hopes and fears surrounding them. But she pushed it all away, focusing solely on the task at hand. Her mind blanked except for the image of the obstacle and the feeling of Whisper beneath her.

The moment of takeoff arrived with the suddenness of a sunrise. There was no dramatic build-up, no sudden burst of energy, a smooth transition, a seamless shift from controlled stride to mighty leap. Leah felt Whisper's muscles bunch and coil beneath her; a reservoir of raw power unleashed in a perfect explosion of controlled energy. They rose, a graceful arc against the backdrop of the hushed arena. For a heartbeat, suspended in the air, the world seemed to fall away, leaving only the vast expanse of sky above and the exhilarating sense of freedom.

The flight was brief, almost ethereal, yet long enough to feel the rush of air against their bodies and the thrill of weightlessness. Leah felt a surge of pure joy, a moment of unadulterated exhilaration as they soared towards the top of their trajectory. It was a moment of pure grace, perfect balance, and profound connection between horse and rider. This culminated in years of training, countless hours of practice, and unwavering dedication, all of which resulted in this breathtaking leap.

The landing was as perfect as the takeoff. Whisper's legs absorbed the impact with effortless ease, her muscles working in harmony to maintain balance and stability. Leah felt a slight tremor, a fleeting moment of uncertainty before the solid ground settled beneath them. The silence that followed was deafening, amplifying the frantic beat of her heart. Then, a wave of relief washed over her, so profound it brought her to tears. They had done it.

The wait felt endless, the arena heavy with tense silence. The judges, their faces impassive, whispered among themselves, their words swallowed by the hush. Leah sat still, trembling, her heart pounding like a drum as a whirlwind of emotions raced through her mind. Whisper, ever the steadfast partner, stood beside her, a silent pillar of strength, her quiet presence a calming balm against the tempest in Leah's soul. The weight of the moment pressed down, heavy and tangible. This wasn't a jump; this was their life's work, and years of dedication poured into this decisive moment.

Finally, the head judge stepped forward. When it came, his voice was quiet yet resonant, carrying across the hushed arena with the authority of finality. With unwavering gaze, he declared, "Leah and Whisper achieved a perfect score, with zero faults."

A sudden sound shattered the previously peaceful silence. As Leah and Whisper basked in the glory of their victory, a tremendous roar from the crowd surged around them like a tidal wave of sound, a powerful testament to their incredible achievement. Cheers, applause, whistles, and joyous cries all blended together to create one enormous, overwhelming wave of approval and admiration from the crowd. Overwhelmed by a potent cocktail of relief, joy, and the unadulterated triumph of victory, tears welled up in Leah's eyes. With gentle hands, she caressed Whisper's neck, her touch light and loving as she murmured words of gratitude and affection into the soft mane of her beloved horse. The victory achieved was not just a simple win, but a powerful demonstration of their unbreakable bond, complete mutual trust, and their collectively shared passion that fueled their efforts.

It validated everything they had worked for, a dream realized, a testament to the power of persistence and the incredible connection between horse and rider. This was more than a competition; it was a journey, a story, a moment etched forever in their shared history. The championship symbolized a bond forged through hard work and triumph as they overcame every obstacle. Both Leah and Whisper would forever remember that soaring leap, the perfect landing, and the crowd's deafening roar. It served as a poignant reminder of their unwavering spirit and their shared victory.

# Chapter 32: Facing the Odds

Although the initial wave of exhilaration had faded, the thunderous roar of the crowd still reverberated in Leah's ears, a persistent, phantom-like sound that lingered long after the event had concluded. As she processed the weight of her accomplishment, the full reality of her victory, the championship title, settled upon her, a feeling of warmth and comfort similar to being wrapped in a cozy blanket. But beneath the surface of exhilaration, a quiet apprehension stirred. Festive celebrations had concluded, marking the end of a joyous period. The lengthy and difficult challenges that she had overcome—the long odds, the intense pressure of the competition, and the relentless self-doubt that plagued her—remained a subtle undercurrent beneath the surface of her triumphant achievement.

The next few days were a whirlwind of activity. Interviews, press conferences, and congratulatory messages flooded in from all corners of the equestrian world. Her phone buzzed incessantly, a symphony of notifications celebrating her victory. Sponsors wanted meetings, journalists clamored for interviews, and fans sent her messages of admiration. It was overwhelming, a tidal wave of attention that threatened to engulf her. Yet, amidst the chaos, Leah found herself strangely detached. The celebratory atmosphere felt surreal, a distant echo of the intense emotions she'd experienced in the arena.

She'd faced long odds, and the victory felt monumental, not because of the competition itself, but because of the internal battles she'd fought and won. Her victory wasn't about skill and technique; it was about overcoming self-doubt, confronting her insecurities, and embracing her vulnerabilities. The pressure had been immense, the expectations almost unbearable. The weight of her family's legacy, the relentless rivalry with Emma, the constant scrutiny of the judges, and the ever-present fear of failure were the obstacles she'd had to overcome, and the scars of those battles remained, visible and tangible beneath the veneer of triumph.

Looking back, she realized the genuine challenge wasn't the height of the jumps, the technical difficulty of the courses, or the fierce competition. The real challenge lay within herself. The long odds she'd faced weren't external; they were internal, psychological barriers she had to conquer. The doubt that had haunted her for so long, the fear that had held her back, had been her most prominent opponent. And by conquering that internal foe, by believing in herself and Whisper, she had achieved something far greater than a championship title.

The victory, however, also amplified the scrutiny. Experts and critics dissected, analyzed, and scrutinized her performances. Every move, every stride, every subtle shift in her posture became a subject of debate and discussion. Leah was the subject of both admiration and criticism, with every action under scrutiny. The pressure to maintain her success and continue her winning streak became immense, a daunting task that threatened to overshadow the joy of her triumph.

Emma's reaction, or the lack of it, was striking. There was no congratulatory message, nor was there any acknowledgement of her victory. Instead, a stony silence had settled between them, a silent tension that underscored the intensity of their rivalry. Emma's absence was more noticeable than any overt display of hostility, leaving Leah with a disquieting sense of unease. Was it grudging acceptance? Or was it simply the prelude to a renewed and even fiercer challenge?

The thought fueled a quiet anxiety within Leah. She'd won, but the battle didn't seem to be over. The competition had ended, but the rivalry continued, unspoken yet palpable, a silent force that hung between them. Leah knew this victory was only a stepping stone, a significant achievement on a long and arduous journey. The pressures of maintaining her success were immense, the expectations even greater. This wasn't about riding but about living up to a legacy, a pressure that weighed on her more profoundly now that she'd succeeded.

In the following competitions, she could test the limits of her recently gained confidence. With pressure mounting and expectations soaring, the situation grew increasingly demanding. Under the blinding spotlight, each jump felt like reliving her final, unforgettable leap at the National Championship. Every stumble and pause drew the judges' scrutiny, magnifying imperfections and leaving no room for error. Previously subdued, the fear of failure returned with a vengeance.

Yet, each time, she rose above it. The lessons learned during the challenging championship had instilled in her a resilience she never knew she possessed. She approached each jump with a new perspective, focusing on the process rather than the outcome. The focus shifted from the fear of failure to the joy of riding, to the connection with Whisper. She learned to harness the pressure, transforming it from a debilitating force into a source of strength and motivation.

The victories that followed weren't about skill and technique; they were about mental fortitude, overcoming self-doubt, and embracing the challenges she faced. They were about reaffirming her belief in herself and Whisper, the unwavering bond that had carried her through the trials and tribulations of the past. Each jump, each competition, became an opportunity to prove her skill, resilience, and ability to overcome adversity and rise to the occasion.

The long odds had made her victory all the sweeter, but it also served as a reminder that every success, no matter how significant, was merely a stepping stone on the path to greater challenges and achievements. The journey wasn't over; it had only begun. She carried the weight of expectation, the legacy of her family, and the relentless pressure of maintaining her success with a newfound grace and determination.

The victory was not a mere win; it was a testament to her resilience, proof of her potential, and a promise of future triumphs. The path ahead was still long and challenging, filled with unpredictable obstacles and relentless pressure. But Leah, armed with the lessons learned and the strength forged in the fires of competition, was ready to face whatever came her way. The long odds had tempered her, strengthened her, and prepared her for the even greater challenges ahead. The journey continued, a thrilling and unpredictable path paved with passion, perseverance, and the unbreakable bond between a rider and her horse.

Leah, ever the ambitious competitor, looked forward to the next jump, the next challenge, the next opportunity to prove that she could once again face the odds and emerge victorious. Rather than a sequence of challenges, the journey served as an ongoing process of growth, refining her character, enhancing her skills, and further solidifying her unyielding spirit. With a future brimming with the exciting prospect of new challenges and victories, Leah felt prepared and eager to embrace whatever destiny had in store for her.

# Chapter 33: A Test of Nerve

The air hung thick with anticipation, a palpable tension that vibrated through the stands and settled heavily on Leah's shoulders. The last jump loomed before her, a formidable obstacle that represented not the culmination of the competition, but a culmination of years of training, relentless self-doubt wrestled into submission, and a fragile hope blossoming into fierce determination. It was a towering obstacle, the final test, the ultimate hurdle between her and victory. The silence before the start was deafening, a stark contrast to the roar of the crowd that had accompanied her every jump thus far. This silence was heavier, more pregnant with expectation, a silence that pressed down on her, demanding an answer, a testament to her courage and skill.

Whisper, her magnificent chestnut mare, stood poised and still beside her, a mirror image of Leah's controlled tension. The mare's breath plumed white in the crisp fall air, a silent testament to the moment's intensity. Their bond, forged in countless hours of training, was palpable; a silent conversation passed between them, a shared understanding of the task at hand. Leah could feel the rhythm of Whisper's breath, the subtle shift of her weight, the quiet power held in reserve, ready to be unleashed at a moment's notice.

The pressure wasn't physical; it was a crushing weight on her mind, a symphony of anxieties playing on repeat. Doubt appeared, suggesting potential failure. Images flickered through her mind—images of past mistakes, falls, and crushing disappointment in her own eyes. She fought against it, pushing back against the tide of self-doubt with every ounce of her willpower. This wasn't a jump; it was a battleground, and her mind was the battlefield.

With a focus on her breathing, she closed her eyes, centering herself in the here and now, actively grounding her awareness. In her mind's eye, she saw herself executing the jump flawlessly, from the perfect approach and trajectory to the effortless landing. As she felt Whisper's comforting weight beneath her, a silent promise of partnership settled upon her, filling her with a sense of calm and security. The unwavering support of her family, a powerful image in her mind, strengthened her determination and bolstered her resolve to continue. Others accompanied her, meaning she was not alone.

With a slow, deliberate movement, she opened her eyes; her gaze settling intently on the challenge that lay before her. The jump tower loomed before her, a seemingly insurmountable and imposing structure constructed of wood and poles; however, within her heart, she experienced a sudden surge of tranquility and a newfound self-assurance that entirely eclipsed her previous apprehension. It wasn't the jump that mattered; rather, it was the arduous journey, the relentless hard work, the unwavering dedication, and the unshakeable belief in herself and her horse that truly defined this achievement.

"It was time to go," Leah gently squeezed and whispered, a silent communication that started their coordinated movement. They began their approach, establishing a measured rhythm between them—a perfect balance of power and control. Whisper responded flawlessly, her stride lengthening and shortening with ideal precision, guided by Leah's subtle cues. The crowd held its breath, their collective tension palpable as they watched the pair approach the jump.

Leah felt a surge of adrenaline coursing through her veins, a thrilling mix of excitement and apprehension. The fear was still there, a faint tremor beneath the surface of her confidence, but it was no longer crippling. It was a source of energy, a reminder of the stakes. This culminated with everything she had worked for; she had to deliver. They were approaching the jump now, gaining momentum, and the final preparations were in perfect synchronicity. The distance was closing; and the jump seemed to grow larger and more imposing. It felt as if time slowed down; the world narrowing to the focus of this single moment, this crucial test of her skill and courage.

Leah focused on every detail: the proper line, the perfect stride, the feel of Whisper beneath her. She didn't think, her body and mind moving in perfect harmony. A profound sense of unity enveloped her, as if she and Whisper were a single entity, moving together toward the challenge.

The moment of truth arrived. Whisper launched herself upward with a powerful surge, soaring effortlessly over the towering obstacle. Leah felt the rush of wind against her face, the thrill of flight, the exhilaration of a perfect execution. There was no hesitation, no stumble; it was a perfect jump.

The landing was flawless. Whisper's four legs touched the ground simultaneously, her muscles absorbing the impact with grace and precision. Leah kept her composure, her body remaining balanced and steady. They moved onwards, not pausing, the graceful continuation emphasizing the perfection of the preceding jump. They continued toward the finish line, a constant rhythm established, Whisper's movements fluid and powerful.

As Leah basked in the moment's glow, a thunderous roar erupted from the crowd, its wave of applause washing over her completely. The weight of their worries lifted, replaced by an immense sense of relief. She had successfully completed the task, achieving her goal. She had conquered not only the last jump but her self-doubt as well. The physical battle equaled the mental one in difficulty. The last jump had been a test of nerve, a crucible that had refined her skill and forged her resolve.

It settled upon her, not as a burden, but as a testament to her hard work, perseverance, and unwavering belief in herself and her horse. It was a moment of pure triumph, a culmination of dreams realized. But even in the face of such overwhelming victory, Leah knew the journey was far from over. This was another stepping stone, a significant milestone, a moment to savor before future challenges beckoned. The future promised new obstacles, challenges, and opportunities to test her mettle. And Leah, with Whisper at her side, looked forward to them all, armed with the strength, skill, and confidence gained from this hard-fought victory.

The awards ceremony was a blur of flashing cameras and congratulatory handshakes. Still, Leah remained grounded, aware that the genuine victory lay not in the external accolades but in her internal transformation. The pressure, scrutiny, and relentless self-doubt were the true obstacles she had overcome. She had faced the long odds and emerged victorious in the competition and the battle within herself.

The rivalry with Emma remained unspoken, a silent tension that hung heavy in the air. Emma's absence from the post-competition celebrations was palpable, a silent challenge hanging like a threat in the distance. Leah knew that their rivalry was far from over, that this victory would only intensify their competition and fuel Emma's determination to reclaim her position at the top. But Leah was no longer intimidated. She had proven to herself that she possessed the skill, the determination, and the mental fortitude to compete at the highest level. The victory had not cemented her place among the elite; it had changed her. She had discovered a wellspring of resilience and strength she never knew she possessed, and this newfound confidence would be her greatest asset in the battles to come.

The memories of the last jump, the test of nerve, would be a constant reminder of her strength, determination, and unwavering belief in herself and her horse. The crowd's roar, the thrill of the jump, and the overwhelming sense of accomplishment would fuel her passion for the sport, driving her toward new heights, new victories, and a future paved with unwavering determination and the unbreakable bond between a rider and her horse.

# Chapter 34: The Execution

The last jump loomed, a stark silhouette against the darkening sky.

It wasn't a jump, but a monument to her journey — a testament to the sweat, the tears, the countless hours spent honing her skills, and the unwavering support of her family and Whisper. The crowd was a hushed sea of faces, their anticipation palpable, their breath held captive in their chests. This wasn't a competition, but a culmination of a lifetime's dedication. This was her moment!

Sensing the shift in the air and the change in Leah's posture, Whisper responded with an almost imperceptible shift in her weight. She stood, a statue of controlled power, her chestnut coat gleaming under the stadium lights. Their connection was a silent language, a shared understanding forged through countless hours spent together, a deep bond woven from trust and mutual respect. Leah could feel the subtle tremor in Whisper's muscles, the anticipation, the shared desire to succeed.

Leah took a deep breath, the crisp fall air filling her lungs. Centering herself, she silenced the incessant chatter of doubt threatening to overwhelm her. She had trained for this moment, visualized it countless times, felt the rhythm of the approach, the power of the leap, the seamless landing. She was one with Whisper, a single entity poised for action.

The approach was a delicate dance of precision and power. Leah guided Whisper with subtle shifts in her weight and the slightest adjustments of her reins. Whisper measured each stride, moved deliberately, and worked every muscle in perfect harmony. The ground rushed beneath them, the distance to the obstacle shrinking with each decisive step. Leah's focus was unwavering, laser-sharp, centered on the jump and the execution of the plan. She blocked out the noise, the cheering, the anxiety, everything except the rhythm of Whisper's gait and the feel of the mare beneath her.

Approaching the jump, the towering obstacle loomed larger, testing her bravery and expertise. But Leah didn't falter. The fear was still there, a subtle tremor beneath the surface of her resolve, but it was no longer the dominant force. It had become a catalyst, a source of adrenaline that sharpened her senses and intensified her focus. The fire fueled her drive, the edge that honed her skill.

The final strides were a blur of motion and perfect coordination. Leah felt the power surge as Whisper lengthened her stride, gathering the momentum needed for the leap. The air hummed with anticipation, the crowd's silence broken only by the rhythmic thud of Whisper's hooves on the ground, a steady drumbeat signaling the approach to the climax. Leah adjusted her position, her body aligning with Whisper's, a seamless extension of the mare's robust frame.

Then, it was time.

With a mighty thrust, Whisper exploded upwards, a breathtaking display of athleticism and grace. Leah felt the exhilarating rush of air against her face, the sensation of flight, and the sheer power of the leap. The world narrowed, focused solely on the jump, Whisper's graceful arc, the precise movement of her muscles, and the unwavering strength of her legs. It was a moment suspended in time, a perfect symphony of movement and skill.

Whisper soared over the obstacle, clearing it effortlessly. Leah's heart pounded in her chest, a mixture of fear and exhilaration. She held her breath, her body tense, but her movements remained controlled, her balance impeccable. Whisper descended, her landing as graceful and precise as her takeoff, her four legs hitting the ground simultaneously. Her robust frame absorbed the shock, and she barely felt the impact.

There was a moment of suspended silence, a collective gasp from the crowd, before cheers erupted. They had done it. They conquered the last jump, the ultimate test. Leah's triumph wasn't about the jump; it was a victory over her self-doubt, over the internal struggles that had plagued her throughout the competition.

They continued their graceful exit. Leah's posture relaxed yet alert, maintaining a perfect rhythm and balance as they made their way to the finish line. The applause echoed around the stadium, a wave of sound that washed over them, a testament to their incredible performance. The crowd's roar was deafening, a tribute to their skill and determination, a validation of years of training and dedication. Leah felt a profound sense of fulfillment, a wave of emotion that washed over her, overwhelming yet exhilarating. She had overcome her doubts and triumphed.

Crossing the finish line marked the end of a goal, more than a physical achievement. Her unwavering spirit, tenacious perseverance, and the unshakeable bond she shared with Whisper culminated in a triumph that served as a testament to her strength and resilience. Their victory resulted from a combined effort, a fusion of two distinct wills, two devoted hearts, and two kindred souls working in perfect synchronization to reach a common goal. A simple rider-horse bond had developed into something much deeper and more meaningful between them.

The judges' scores confirmed Leah's flawless execution, and the championship title felt like a badge of honor, a tangible representation of her achievement, rather than a burden. The accolades that followed were a blur of flashing lights and congratulations, a whirlwind of excitement. Still, for Leah, the valid reward was the feeling of accomplishment, the sense of overcoming her self-doubt, the deep satisfaction of having achieved something remarkable.

She dismounted, carefully easing Whisper to the ground, her touch gentle, her words softly whispered in praise. The mare responded with a soft whinny, a grateful sigh. Their shared victory was more profound than words could ever describe. It was a testament to the power of dedication, the reward of years of perseverance, and the strength of the bond between horse and rider.

The post-competition celebrations were a whirlwind of activity, a blur of congratulations and celebrations. But Leah remained grounded, aware of the significance of this victory, not just in terms of the championship title, but also in her personal growth, self-discovery, and the evolution of her relationship with Whisper. She had not only conquered the jump, but also her fears and insecurities.

They noted Emma's absence, and also a silent undercurrent in the joyful atmosphere. The unspoken rivalry remained, a potent force that simmered beneath the surface. Leah knew that the competition was far from over, that Emma would wait in the wings, ready to challenge her supremacy. But Leah was no longer intimidated. She had proven to herself that she possessed the skill, resilience, and courage to compete at the highest level.

# Chapter 35: The Outcome

The final rail fell. A collective gasp rippled through the crowd, followed by an almost deafening silence, broken only by the gentle thud of Whisper's hooves as she landed. The silence stretched, a taut string pulled to its breaking point, before exploding into a thunderous roar of applause and cheers. They had done it. Zero faults.

Still perched atop Whisper, Leah felt a wave of relief so intense it almost knocked her off balance. Her heart hammered against her ribs, a frantic drumbeat against the echoing cheers. The fear, the self-doubt, the crushing weight of expectation all melted away, replaced by a euphoric rush of adrenaline and pure joy. She had done it. She had conquered the last jump, not the physical obstacle, but the internal demons that had haunted her throughout the competition.

Sensing Leah's emotional shift, Whisper nuzzled her neck softly, her warm breath comforting against her cheek. The mare seemed to share in the victory, her body trembling slightly with the residual energy of the jump, yet her posture remained steady, proud, unwavering. They were a team, a single entity that had overcome seemingly insurmountable odds.

The exit from the ring felt surreal. The world blurred into a kaleidoscope of flashing lights and blurred faces, a torrent of sound that washed over them, yet somehow didn't penetrate the bubble of shared triumph that enveloped Leah and Whisper. She could feel the vibrations of the crowd's energy beneath her, a physical manifestation of their collective elation.

As they approached the in-gate, the cheers intensified, reaching a crescendo as they stepped out of the arena. Leah felt a strange detachment, a sense of watching herself from a distance, as if she were observing a movie rather than experiencing her own life. The reality of her victory still hadn't fully sunk in.

Officials announced the scores as a formality, merely confirming what they already knew: zero faults. The championship was theirs. The weight of the title didn't crush her; it lifted her, buoying her with a profound sense of accomplishment. She dismounted, carefully guiding Whisper to the ground, showering her with gentle pats and whispered words of praise. The mare responded with a soft whinny, a quiet expression of shared joy and relief.

The subsequent flurry of activity was a whirlwind of activity. Photographers flashed their cameras, capturing images of the triumphant pair. Fellow competitors, trainers, and officials offered their congratulations, their voices a jumbled chorus of celebratory greetings. With expressions of profound pride and unmistakable relief, her family members surged forward as one. Overwhelmed with emotion, her father embraced her tightly, his eyes shining, and the tears streaming down his face muffled his words of comfort. Her mother, a woman typically known for her composure and reserved nature, unexpectedly and completely broke down into tears of unadulterated joy, her face streaming with them. Even her younger brother, usually reserved, clapped and cheered, his face lit up with genuine excitement.

Leah soaked it all in, each moment a precious memory etched into the tapestry of her life. But amidst the chaos and whirlwind of congratulations and celebrations, she sought the quiet moments, the intimate exchanges with Whisper. They shared a glance that spoke volumes more than words ever could. Countless hours of training and unwavering support tested their bond, and it emerged stronger, deeper, and more resilient than ever before.

The post-competition celebrations comprised champagne toasts, celebratory dinners, and endless congratulations. Leah, however, found herself increasingly drawn to the quiet solace of the stables. She spent hours grooming Whisper, their connection deepening with each brush stroke. The mare seemed to understand her need for quiet contemplation, her presence a calming influence amidst the excitement.

In the quiet of the stables, away from the clamor of the festivities, Leah had time to reflect on her journey. The competition had been more than a test of her riding skills; it had been a crucible that forged her character, strengthening her resolve and deepening her understanding of herself. She had confronted her self-doubt, faced her fears head-on, and emerged victorious.

The victory was bittersweet. While she savored the triumph, a quiet part still carried a sense of unfinished business. Emma's absence hung heavy, a silent reminder of the remaining unspoken rivalry. Leah knew that Emma's absence wasn't a sign of surrender, but a temporary retreat — a regrouping before the next challenge. The competition, Leah knew, was far from over. But this time, quiet confidence — a newfound belief in her abilities replaced the uncertainty. She had proven to herself, not others, that she was a force to be reckoned with.

The championship title acknowledged her effort and commitment. It was a testament to the countless hours spent honing her skills, her family's unwavering support, and the deep bond she shared with Whisper. However, the most valuable reward was the internal transformation and self-discovery that accompanied her victory. She had won a competition and a battle against her self-doubt, emerging stronger, more confident, and more resilient than ever before.

The road ahead wouldn't be easy. She knew future competitions would present new challenges, new obstacles to overcome. But she faced the future with a renewed sense of purpose, armed knowing that she possessed the skill, the determination, and the unwavering support of Whisper to conquer whatever lay ahead. This victory was not an end; it was merely a stepping stone, a testament to her journey, a prelude to even greater triumphs. The thunderous applause of the crowd lingered in her memory, a powerful reminder of her accomplishment and a wellspring of motivation as she faced the difficulties that lay ahead.

Although the future held countless challenges and obstacles, she met them not with fear or trepidation, but with the quiet and unwavering confidence of a champion who knew, deep within her heart, that she possessed the strength, skill, and determination to overcome every one of them. With every victory, she not only grew and learned, overcoming her limitations, but also deepened the unshakeable bond she shared with her devoted companion, Whisper. Their partnership was more than a collaboration; it was a testament to the unwavering spirit of the horse and rider, a collaboration of will and skill that would conquer any obstacle.

# Chapter 36: Announcing the Results

The loudspeaker crackled to life, amplifying the voice across the hushed stadium, a stark contrast to the previous roar of the crowd. A hush fell over the assembled spectators, a tangible wave of anticipation that pressed down on Leah, even more intense than the pressure she'd felt during the last jump. She stood beside Whisper, the mare calm and seemingly unfazed by the charged atmosphere, a comforting weight against her leg. The silence stretched, a pregnant pause that seemed to amplify the beating of her own heart. It felt as though time itself held its breath, waiting.

The announcer's voice, crisp and clear, cut through the stillness. He began with the lower placings, meticulously detailing each rider's performance, each score, each fence navigated, each rail that fell. With each name, a ripple of murmurs stirred through the crowd, a mixture of congratulations and commiseration. Leah barely registered the names, her focus entirely on the moment and the impending announcement of her score. Her body felt strangely numb, yet intensely alert, her senses hyper-aware of every nuance in the atmosphere. The expectant silence magnified even the faintest rustle of a program or the quiet cough of a spectator.

Finally, the moment arrived. The announcer cleared his throat, drawing the silence tighter, the anticipation reaching a fever pitch. He spoke the words Leah had been waiting for, the words that would solidify her victory, or shatter her hopes. "And in first place," he announced, his voice resonating with a gravity that mirrored the moment's importance, "with a score of zero faults, is... Leah Holloway and Storm's lil Whisper!"

The silence that had held its breath shattered. An explosion of cheers erupted, and a wave of sound washed over Leah, engulfing her in a sea of joyous noise. Confetti rained down from the stands, a vibrant shower of color that swirled around her and Whisper. The world seemed to tilt on its axis, the reality of her victory crashing down upon her with the force of a tidal wave. This time, it wasn't the detached observation of the earlier moments, but pure elation. The years of hard work, dedication, sweat, tears, and self-doubt all culminated in this single, glorious moment.

The crowd surged forward, a sea of faces, all beaming with smiles and congratulations. Leah felt warmth wash over her, a collective embrace of support and recognition. The sheer volume of sound was overwhelming, yet strangely comforting, a testament to the achievement celebrated by her and the entire equestrian community. She felt a profound sense of gratitude, not for her skill and hard work, but for the support of her family, her trainers, and, most importantly, Whisper.

With a mixture of tearful embraces and joyful shouts, her family rushed toward her, their emotions a palpable wave of relief and happiness as they enveloped her in a comforting embrace. Overwhelmed with pride, a smile creasing his face and revealing the depth of his paternal affection, her father embraced her in a hug so tight that she found it difficult to breathe. Usually a picture of quiet reserve. Her mother wept openly, tears of pure, unbridled joy flowing freely down her face. Even her typically quiet and reserved younger brother, surprisingly, clapped and cheered, his face shining with an enthusiasm that clearly reflected his excitement about the event. They were her anchor, her support system, and in this moment of triumph, she felt an overwhelming sense of gratitude for their unwavering belief in her.

The photographers swarmed around them, flashing their cameras, capturing the moment for posterity. The whirlwind of emotion and celebration enveloped Leah. Yet, amidst the chaos, she found herself drawn to the quiet moments and the intimate exchanges with Whisper. She leaned down, burying her face in the mare's soft mane, whispering words of praise and gratitude. Whisper responded with a soft whinny, her breath warm against Leah's cheek. It was a silent communication that transcended words. It was a moment of shared triumph, a testament to their unbreakable bond.

The post-competition celebrations were a blur of activity. There were champagne toasts, celebratory dinners, interviews with journalists, and endless congratulations from fellow competitors, trainers, and officials. Each interaction felt surreal, a dreamlike sequence of events that seemed to float outside the realm of reality. Yet, amidst the excitement, Leah found herself repeatedly drawn back to the quiet solace of the stables.

She spent hours grooming Whisper, their connection deepening with each brush stroke. The mare seemed to understand her need for quiet contemplation, her presence a calming influence amidst the excitement. In the quiet of the stables, surrounded by the familiar scent of hay and leather, Leah allowed herself to reflect on her journey, which had brought her to this moment of triumph.

The celebrations continued for days, a whirlwind of joyous activity that left Leah elated and exhausted. But amidst the excitement, a sense of quiet contemplation settled over her. She spent her time reflecting on the competition, the challenges she had overcome, and the lessons she had learned. The absence of Emma weighed on her mind, a subtle reminder of the unspoken rivalry that still lingered between them. Emma's absence wasn't a sign of defeat, Leah knew, but a strategic retreat, a regrouping before the next challenge. As she reflected on the competition, she realized the end was still a long way off.

Although she had won, the victory ignited within her a quiet confidence, a new fire that radiated outwards, a beacon of quiet strength. Winning the championship title served as a powerful validation, conclusively confirming the extent of her exceptional skills and considerable abilities. The achievement served as powerful evidence of the countless hours of rigorous training she had undertaken, the unwavering support provided by her family, and the profound, unshakeable bond that existed between her and her horse, Whisper. But the most precious prize, the valid reward, was the self-discovery — the transformation that had taken place within her throughout the journey. She had not merely won a competition; she had won a battle against her self-doubt, emerging stronger, more confident, and more resilient than ever before.

She realized the road ahead would be difficult. Upcoming competitions will pose fresh challenges and obstacles. However, with newfound purpose, skill, determination, and Whisper's unwavering support, she bravely faced the future and whatever challenges it held. This victory didn't conclude; In her heart, the echoes of the crowd's roar still resonated, a constant reminder of her achievement and a wellspring of inspiration for challenges to come. The journey had only begun, and Leah, with Whisper by her side, felt ready to ride. The future held countless jumps.

Now, though, she met their gaze not with apprehension, but with the self-assuredness of a champion who knew, deep down, she could defeat them all. Each victory fueled her growth, learning, and overcoming of limitations, strengthening her unbreakable bond with Whisper. Their partnership developed beyond a simple working relationship, developing into something far more profound and significant. Despite winning the championship—a significant achievement—their journey pressed on. The sweetness of victory foreshadowed even greater rewards. Despite the uncertainty of the future, the journey and adventure were imminent.

# Chapter 37: Leahs Reaction

The initial wave of elation, the sheer, overwhelming joy of hearing her name, announced as the champion receded, leaving behind a quiet hum of disbelief. It wasn't the confetti, the cheers, or the celebratory hugs that felt surreal; it was the internal shift, a profound change in her perception of herself. For years, the shadow of her family's equestrian legacy had loomed large, casting a daunting shadow over her ambitions. She'd felt the weight of expectation, the pressure to succeed, a pressure that often manifested as crippling self-doubt. The fear of failure had been a constant companion, a silent specter that haunted her every ride. But now, standing amidst the cacophony of celebration, that fear was strangely absent. She hadn't erased it, but the fear felt muted, softened, almost manageable.

The celebratory dinner felt like a scene from a dream. The Champagne tasted fizzy and sweet, and her family's laughter was a comforting melody against the background buzz of conversation. Her father, usually stoic and reserved, was beaming, his eyes shining with a pride that warmed her heart. Her mother, traditionally composed and elegant, openly wept, her tears a testament to the years of shared anxieties and hopes. Even her younger brother, usually more interested in video games than horses, was engaged in lively conversation, proudly boasting about his older sister's win to anyone who would listen.

Throughout the evening, Leah drifted away from the throng, seeking moments of quiet solitude. The constant barrage of congratulations, while heartfelt, was overwhelming. She needed moments to process the enormity of her victory, to fully absorb the significance of what she had accomplished. She gravitated towards the stable, the familiar scent of hay and leather, and the calming presence of Whisper. The mare, seemingly unfazed by the flurry of activity surrounding her, stood patiently in her stall, her breath slow and even.

Leah spent a long grooming on her, the repetitive motion a soothing balm to her frayed nerves.

The touch of Whisper's soft coat, the rhythmic brushing, helped ground her and reconnect her to the tangible reality of her achievement. In those quiet moments, far removed from the bright lights and boisterous celebrations, Leah truly felt the weight of her victory. It wasn't about the zero faults, the perfectly executed jumps, or the coveted championship title. It was about the journey, the years of relentless training, the setbacks, the self-doubt, the countless hours spent honing her skills, nurturing her talent, and building an unbreakable bond with her horse. The championship marked the culmination of that journey, a tangible manifestation of her dedication and perseverance.

This victory felt different from any she had achieved before. Previous wins had brought a sense of accomplishment, a fleeting surge of pride, but a nagging undercurrent of self-doubt had always accompanied them. She'd often questioned whether she had truly deserved the win, whether it was a matter of luck rather than skill. But this time, she lessened her self-doubt, replacing it with quiet, deep-seated confidence. She had faced her inner demons, wrestled with her insecurities, and emerged victorious. The victory wasn't about beating her competitors, but about conquering her self-limiting beliefs.

Her championship ribbon symbolized not only her riding prowess, but also her personal growth, a testament to resilience and determination. She spent the following days in interviews, photo shoots, and media appearances. The attention was both exciting and overwhelming. Leah navigated a whirlwind of press conferences, answering questions about her strategy, training regimen, and feelings about the competition. She handled the interviews with poise and a newfound confidence, and a calm demeanor and articulate responses were clear in her responses. It was as if the victory had given upon her a championship title and a newfound eloquence, a gift of self-assurance that enabled her to communicate her thoughts and feelings with clarity and conviction.

However, a quieter reflection was taking place beneath the surface of excitement and the whirlwind of media attention. Leah analyzed her performance, identifying areas for improvement, and planning for future competitions. The victory had not made her complacent; instead, it had ignited a renewed passion, a more profound commitment to her sport. She knew rigorous training, tireless dedication, and a constant pursuit of excellence paved the path to continued success.

Emma's absence from the awards ceremony and post-competition celebrations was keenly felt, a silent presence lingering in the background. Though Leah initially savored her victory, the tension of their competitive relationship simmered just beneath the surface.

Leah realized that Emma's absence wasn't a sign of defeat; it was a calculated move, a strategic retreat, a period of reflection and regrouping before the subsequent encounter. Leah knew that their rivalry was far from over, and her quiet understanding of this subtle power play added another layer of complexity to her triumph.

It was in the quiet moments, in the solitude of the stable, that Leah truly expressed her emotions. Joy and responsibility intertwined as she felt the weight of maintaining her performance and pushing for even greater achievements. The pressure was still there, but it no longer felt oppressive. It was now a motivating force, a challenge to be embraced rather than feared. She had overcome her self-doubt but recognized that self-improvement was an ongoing process, a continuous journey rather than a single destination.

As the initial euphoria subsided, a more profound sense of gratitude settled over Leah. She felt a profound appreciation for her family's unwavering support, her trainers' guidance, and the steadfast partnership with Whisper. It had been a team effort — a collective journey toward a shared goal. The victory was not solely hers; it belonged to everyone who had played a role in her success.

The championship title validated her hard work and dedication, but it also served as a stepping stone, a launchpad for future achievements. Leah knew the journey was far from over; the road ahead was likely filled with challenges, setbacks, and new obstacles to overcome. But she faced the future with a renewed sense of purpose, armed with a champion's quiet confidence, which emanated from her core.

The whispers of doubt that had once haunted her were now muted, replaced by a quiet inner strength. She had won a championship and battled her insecurities, transforming her from a talented young rider plagued by self-doubt into a confident, resilient champion. This transformation was perhaps the most valuable prize, a personal victory far more significant than any trophy or title. The future stretched before her, a limitless expanse of opportunities, and with Whisper at her side, she was ready to ride.

# Chapter 38: Emma's Response

E mma's silence was deafening. There was not a single congratulatory message, and there was no acknowledgment of Leah's stunning victory. Leah had expected some response, perhaps a grudging nod of respect, even a veiled threat for future competitions. But nothing. The absence starkly contrasted with the celebratory whirlwind surrounding Leah, a void that felt almost as significant as the victory itself. It wasn't the lack of a message; it was the palpable shift in the atmosphere, the subtle change in how other riders interacted with Leah — a noticeable shift in the hierarchy of the equestrian world. Before, there had been a palpable tension, a silent acknowledgment of their rivalry, a mutual respect laced with unspoken animosity. Now the tension remained, but the dynamic had altered. Emma's silence was strategic, a calculated move designed to unsettle, to sow seeds of doubt in the minds of those who were now looking at Leah with fresh eyes.

Leah analyzed Emma's absence. Was it pride wounded, a refusal to acknowledge defeat? Or was it something more strategic, a calculated retreat designed to regroup and return with vengeance? Emma was a formidable competitor, her skill undeniable. She didn't accept defeat easily. With a chilling certainty, Leah knew this wasn't the end. This was another chapter in their ongoing rivalry, a lull before the storm. Their silent competition had become a complex, unspoken dance, each move calculated, each step deliberate, each silence pregnant with meaning.

The weeks following the championship were a blur of training, media appearances, and sponsorships. Leah's win had catapulted her into the spotlight, transforming her from a promising young rider to a rising star. The pressure, however, had not lessened. The expectations were higher, the competition fiercer, and the stakes significantly greater. She scrutinized her every move, second-guessing every decision. The weight of expectation was now testing the quiet confidence she had seen in the aftermath of her victory. To consistently deliver faultless performances, she had to stay on top. The fear of regression, of falling short of expectations, was a constant threat, a shadow lurking in the corners of her mind.

She sought guidance from her trainer, a veteran equestrian with years of experience in high-stakes competitions. Praising her win, he reminded her that victory was not a destination, but an ongoing process. He stressed the importance of constant refinement, continuous improvement, and always pushing her limits. He warned her against complacency, urging her to remain hungry, to strive for perfection, even in the face of overwhelming success. His words were a harsh reminder that the equestrian world was unforgiving; one slip-up could send her down the ranks.

The intensity of her training regime was relentless. She spent hours honing her technique, refining her skills, and strengthening her bond with Whisper. The mare responded to Leah's determination with unwavering partnership, her movements becoming increasingly fluid and her jumps flawless. But the pressure was taking its toll. Leah's shrinking social life and disrupted sleep meant she spent less time with her family. She was hyper-focused on her goals, her entire life revolving around training, competition, and the constant pursuit of excellence.

Then, one evening, she received an email after a grueling training session. It was from Emma. Not a congratulatory note, not an apology, but an invitation. An invitation to a private training session at Emma's exclusive stables, known for its exceptional facilities and intense training regimens. The email was brief, almost curt, devoid of any pleasantries. It simply stated the date, time, and location. There was no explanation, no context, a stark invitation that sent a shiver down Leah's spine.

Leah wrestled with the decision. A part of her was hesitant, wary of Emma's motivations. What did she want? Was this an olive branch, a gesture of camaraderie? Or was it a trap, a strategic maneuver to study Leah's techniques, to uncover her weaknesses? The invitation was a calculated risk, a move that could either strengthen their rivalry or create an unexpected alliance. The implications of accepting or rejecting the invitation stretched far beyond a simple training session; it was a move that could redefine their relationship, alter the dynamics of the competition, and potentially shape the trajectory of their equestrian careers.

The uncertainty gnawed at her, keeping her awake at night. She replayed their interactions in her mind, analyzing Emma's expressions, interpreting her silence, and deciphering the subtext of their unspoken communication. Emma's enigmatic nature was a source of both fascination and apprehension. She was a formidable competitor and an enigma—a complex character who concealed her true intentions beneath a veil of calculated aloofness.

After days of internal debate, Leah accepted the invitation. She wouldn't allow Emma to dictate her actions or control her reactions. Instead, she would confront the uncertainty, face the unknown, and engage with the challenge head-on. She would not let fear paralyze her or doubt deter her. Her success in the previous competition gave her newfound resilience and confidence beyond the equestrian arena.

The training session was intense. Emma pushed Leah to her physical and mental limits, relentlessly scrutinizing her technique, demanding precision and perfection. There was no camaraderie, no pleasantries, a focused, relentless pursuit of excellence. It was a grueling test of skill, stamina, and mental fortitude, a stark contrast to the celebratory atmosphere of the championship. However, it was during that intense training that Leah felt Emma's quiet respect. Emma didn't speak or declare it, but Leah felt her respect. The subtle shift in Emma's demeanor — a softening of her gaze, a less pronounced air of arrogance — spoke of grudging admiration.

An unspoken understanding concluded the session. The rivalry had not yet ended. It wasn't quite friendship; rather, a reluctant partnership born from their mutual passion, a temporary peace in their equestrian rivalry. The post-training silence was unlike before. This was a moment of silent respect. The silence spoke volumes, a mutual respect between tough rivals.

# Chapter 39: Acceptance and Grace

The championship aftermath wasn't about celebratory dinners and overflowing inboxes. It was about the quiet moments, the introspection, the subtle shifts in dynamics that unfolded like a slow-motion replay of the competition itself. The media frenzy, the sponsors clamoring for her attention, the endless interviews–it all swirled around her, a whirlwind of noise that threatened to drown out the quiet hum of her thoughts. She had won, undeniably, but the victory tasted different from what she'd imagined. It wasn't the sweet, unadulterated triumph she'd envisioned. There was a lingering undercurrent of unease, a subtle dissonance that echoed Emma's silence.

One evening, while reviewing footage of the final round, she noticed something she'd missed in the heat of the moment. A slight hesitation in Whisper's stride approaching the last obstacle, a subtle shift in her weight that only Leah, intimately familiar with her mare's every nuance, could detect. It was barely perceptible, a fleeting imperfection in an otherwise flawless performance. Yet, it was enough to make her pause. Had she pushed Whisper too hard? Had she overlooked a crucial detail in her training? The nagging doubt burrowed into her mind, a persistent whisper contrasting with the public accolades.

This self-reflection, this critical examination of her performance even in the face of victory, marked a turning point. It was a testament to her growing maturity, recognizing that perfection wasn't about winning; it was about striving for continuous improvement, acknowledging imperfections, and learning from them. This wasn't simply about honing her riding skills, but about cultivating a profound self-awareness that would serve her inside and outside the equestrian arena. She realized that true grace wasn't about flawless execution but the ability to accept triumph and setbacks equally.

Her trainer, a grizzled veteran who had seen countless champions rise and fall, understood this implicitly. Naturally, he praised her victory, but he also stressed the value of humility. The fleeting nature of victory was a topic of his discussion. He advised her to focus on consistent skill improvement, her bond with Whisper, and mental fortitude, rather than dwelling on her win. He noted setbacks are inevitable, even for the most seasoned professionals.

This newfound perspective influenced her approach to training. It wasn't about pushing harder, faster, striving for unattainable perfection; it was about listening to Whisper, understanding her limitations and strengths, fostering a partnership built on trust and mutual respect. She incorporated mindfulness techniques into her routine, focusing on breath control and mental visualization, which helped her manage the pressure and stay centered during intense training sessions. This wasn't about physical prowess, but mental agility and emotional intelligence.

The invitation from Emma, initially a source of trepidation, now felt less like a threat and more like an opportunity for growth. Leah chose not to dwell on the underlying motivations, the potential for manipulation or sabotage. Instead, she saw it as an opportunity to observe a different approach, analyze a distinct training style, and learn from a formidable competitor. She recognized that the rivalry, while intense, could also serve as a catalyst for her improvement.

The training sessions with Emma were intense, demanding, and ultimately revealing. Emma pushed Leah to her limits, relentlessly scrutinizing her technique, requiring pinpoint accuracy and impeccable timing. There were no pleasantries, no casual conversation; it was a raw, focused exchange of skill and determination. But within that intensity, something unexpected emerged. Emma, known for her icy demeanor and intimidating presence, showed glimpses of something akin to respect. The respect wasn't overt or effusive; It was in these shared moments of intense focus, in the mutual understanding of the physical and mental demands of their sport, that a grudging respect blossomed. It wasn't friendship, not camaraderie, but something far more profound–a mutual appreciation for their shared passion, a silent acknowledgment of each other's talent and determination. They were still competitors, locked in a fierce rivalry, but their relationship was developing, transcending the simple framework of winner and loser. It became a quiet understanding, an unspoken truce forged in the crucible of their shared ambition.

Leah learned to embrace victory and defeat, not as opposites, but as integral parts of a continuous journey. True success, she understood, wasn't just trophies and ribbons, but personal growth, unwavering commitment, and quiet resilience developed through challenges. She learned to balance ambition and humility, recognizing her flaws while striving for excellence. She realized that even the most formidable competitors could, in unexpected ways, become sources of inspiration and learning. And above all, she learned the importance of grace, both in victory and defeat–the ability to accept the outcomes with dignity, to learn from setbacks, and to continue striving for excellence with unwavering determination.

The actual test was not the jumps, but how she handled everything around them. And Leah was ready for whatever came next.

# Chapter 40: Lessons Learned

A tranquil rhythm replaced the frenetic energy of the championship; the post-championship days were markedly different. Awards, interviews, and cameras all blurred, making way for deeper thought. What was initially pure elation, the victory now felt more multifaceted, less simple. The blue ribbon wasn't the point; Leah realized, wasn't a destination but a stepping stone. It was a validation of her hard work, her dedication, and her unwavering commitment to her craft.

However, it was also a stark reminder of the constant need for improvement and the relentless pursuit of excellence. The slight hesitation she'd detected in Whisper's stride during the final round became a microcosm of her entire journey. That imperceptible flaw, hidden within the brilliance of their performance, served as a potent reminder that even in success, there's always room to grow. Her relationship with Whisper deepened beyond the arena. The horse, her steadfast partner, had become more than a mount; she was a confidante, a source of strength, a mirror reflecting Leah's emotional landscape. Whisper's subtle cues, previously overlooked in the pressure of competition, now held a profound significance.

Leah honed her understanding of her mare through careful observation, noticing subtle shifts in behavior and energy. This heightened awareness extended beyond horses, enhancing her relationships and attuning her to the nuances of nonverbal communication. The repercussions extended beyond her relationship with Whisper. Fierce competition, performance pressure, and high expectations forced her to address her weaknesses and insecurities. Show jumping lessons went beyond the technical aspects. Leah had made significant strides in her strategic understanding.

Analyzing the performance of other riders, observing their techniques, and studying their approaches had broadened her perspective. She realized that success wasn't about individual skill but strategic planning, adaptability, and the ability to expect challenges. The meticulous planning and preparation she'd invested in, initially driven by a need to prove herself, had become a source of empowerment. She now saw meticulousness in controlling and accepting what she couldn't.

The competitive arena did not encompass the entire transformation. The quiet moments of self-reflection, the insightful conversations with her trainer, and the unexpected connection with Emma—these experiences had fostered a profound sense of self-awareness and emotional intelligence. Leah recognized the importance of balance and integrating her passion for equestrian sports with other aspects of her life. She understood that true fulfillment lay not in achieving external success, but also in cultivating inner peace and harmony. The newfound respect for Emma was impactful. Fueled by mutual ambition and a healthy dose of competitive spirit, the initial rivalry had morphed into something far more nuanced.

Initially perceived as a formidable opponent, Emma revealed unexpected layers of depth and complexity. Their shared passion for the sport, dedication to excellence, and mutual understanding of the challenges forged a surprising connection. It wasn't friendship, but a mutual appreciation for the discipline and commitment required to excel. It was a lesson in seeing beyond superficial judgments and recognizing shared values and goals.

This recognition of shared humanity extended to her interactions with others within the equestrian community. She learned to appreciate her family's support, her trainer's guidance, and the camaraderie of fellow competitors. We initially saw competition as a win-lose situation, but now we see it as something we can all work together on. Leah realized the shared passion for the sport could foster strong bonds and valuable relationships, even amid fierce competition. It was a valuable teamwork and collaborative learning lesson, even in a field often dominated by individual achievements.

Her self-discovery wasn't merely about personal growth within the confines of the show jumping world, but about realizing the transferrable skills that riding instilled in her. The discipline, perseverance, and dedication required by competitive riding permeated every aspect of her life. The ability to manage pressure, maintain focus under intense scrutiny, and bounce back from setbacks proved invaluable in other areas. These skills built resilience in her academic pursuits, personal relationships, and daily problem-solving approach.

The immediate aftermath of the competition did not contain all the lessons learned. They continued to shape her training regimen, approach to competitions, and overall perspective on life. The process of continuous improvement, pursuing excellence, and the balance between ambition and humility–these lessons became integral parts of her daily routine, not in riding but in everything she undertook.

The lessons she learned during the show jumping competitions had a profound and lasting impact, extending far beyond the arena itself, shaping not only her character but also her overall approach to tackling life's inevitable difficulties. The true reward wasn't the physical trophy itself, but the profound and transformative personal journey that achieving it represented, a journey of growth and self-discovery. Leah knew, with absolute certainty, that this victory, symbolized by the soon-to-fade ribbons, would remain a cherished memory long into the future.

# Chapter 41: Post-Competition Adjustments

The days that followed the competition were a strange blend of quiet satisfaction and restless energy. The adrenaline rush had subsided, leaving behind an exhilarating and unsettling calm. Leah found herself adrift in a sea of post-competition introspection, the echoes of cheering crowds and the clang of hooves on the arena floor replaced by the gentle rhythm of Whisper's breathing in the stable. The victory felt distant, unreal, like a vivid dream she might soon forget. Yet, the tangible weight of the blue ribbon resting on her dresser was a constant reminder of her accomplishment.

She spent hours in the stable, meticulously grooming Whisper, the repetitive strokes soothing her still-racing mind. Each brushstroke was a silent conversation, a way of connecting with her equine partner, of thanking her for their shared triumph. Whisper, sensing Leah's quiet contemplation, nuzzled her hand, her soft breath a gentle reassurance. Their bond, forged through competition, had grown deeper than ever. It wasn't the physical connection, the seamless understanding between rider and horse; it was a more profound empathy, a shared sense of pressure and triumph, vulnerability, and strength.

The post-competition analysis began subtly, a quiet internal dialogue that gradually blossomed into a full-fledged review. She reviewed the videos of her rounds, dissecting her performance frame by frame, identifying areas for improvement, even in the winning runs. The slightest hesitation in her approach to the obstacle, a fraction of a second's delay in her release over the triple bars—these microscopic flaws, invisible to the casual observer, became the focal points of her self-assessment. It was a rigorous self-critique, fueled not by self-doubt but by a relentless drive for perfection.

Her trainer, Mr. Henderson, a seasoned veteran with an eagle eye for detail, provided invaluable guidance. His feedback wasn't about dwelling on past mistakes, but about strategically planning for the future. Helped her identify specific areas to focus on in her training, refining her approach to the more challenging jumps, strengthening her control over Whisper's pace, and enhancing their overall communication. The training sessions became a structured process of refinement, a meticulous sculpting of their performance, aimed at eliminating the subtle imperfections that even a champion performance contained.

The conversations with Mr. Henderson extended beyond the technical aspects of show jumping. They delved into the mental game, the importance of maintaining focus under pressure, and the strategies for managing anxiety. He introduced her to mindfulness techniques, suggesting practices to help her center herself before and during competitions, to manage her nerves and focus on the task at hand. Initially unfamiliar and slightly awkward, these techniques gradually became an integral part of her pre-competition ritual, providing a sense of calm and control in the face of mounting pressure.

Leah's perspective on competition also shifted subtly. It wasn't merely about winning; it was about continuous improvement, pushing personal boundaries, and learning from triumphs and setbacks. The initial euphoria of victory had given way to a more grounded appreciation for the process, the challenges overcome, and the lessons learned along the way. She viewed competition not as a battle against her opponents, but as a collaborative pursuit of excellence, a shared challenge that fostered personal growth and mutual respect.

This newfound perspective led her to re-evaluate her relationship with Emma. The competitive edge that had once defined their interactions softened, replaced by a more nuanced understanding. They shared insights, exchanging tips and techniques without the undercurrent of rivalry. Once perceived as arrogance, Emma's confidence now appeared as self-belief, a strength that Leah admired. Their discussions expanded beyond the competitive arena, delving into the everyday challenges of balancing athletic pursuits with academic demands, managing the expectations of family and friends, and navigating the emotional highs and lows of a demanding sport.

Beyond Emma, Leah broadened her circle of camaraderie within the equestrian community. She actively sought advice from more experienced riders, observing their techniques and learning from their experiences. She engaged more readily with fellow competitors, exchanging stories and insights, fostering a sense of collective effort rather than isolated competition. In an unforeseen and positive turn of events, the community's sense of togetherness offered a surprisingly high level of support and personal development.

Her equestrian training imparted valuable lessons that proved to be far-reaching, significantly affecting areas outside of the equestrian sports world. Through the rigorous discipline and unwavering perseverance demanded by competitive riding, she cultivated a powerful work ethic that not only improved her academic performance but also significantly enhanced her focus and overall ability to achieve her goals across various aspects of her life. Faced with intense scrutiny, the skill of managing stress and pressure proved essential to overcoming the difficulties inherent in schoolwork, the nuances of social interactions, and the challenges often presented by family dynamics. Her experiences on the course, though marked by setbacks and challenges, unexpectedly strengthened her resilience; this hard-won resilience fueled a persistent determination in her life, empowering her to overcome any obstacle she faced with remarkable tenacity.

Not only did her competitive edge improve, but her newfound maturity and self-awareness also influenced many other areas of her life. She added regular exercise to her daily schedule because it provided not only physical benefits but also improved her mental clarity and fostered a sense of peace and tranquility. Friends and family received a greater share of her time, as she carefully cultivated relationships that blossomed into a rich and rewarding life that extended far beyond the confines of the equestrian arena. Recognizing the significance of incorporating periods of rest and reflection into her routine, she prioritized downtime to prevent burnout and maintain a sustainable pace in pursuing her passion.

The post-competition period, therefore, wasn't simply a time of relaxation and celebration; it was a period of profound personal growth and transformation. It was a time of reflection, self-assessment, and strategic planning for the future. Leah emerged from the competition as a champion and a more resilient, self-aware, and emotionally intelligent young woman. The lessons learned, both on and off the course, laid a firm foundation for her future endeavors, ensuring that her journey in the world of competitive show jumping would be as much about personal growth as about athletic achievement.

The victory was significant, but the transformation was far more profound. And Leah knew, with a quiet certainty, that her journey had begun.

# Chapter 42: Developing Relationships

The quiet hum of the stable, usually a comforting backdrop to Leah's life, felt charged with unspoken emotions in the days following the competition. The victory, once a dazzling spotlight, now cast long shadows, highlighting the complexities of her relationships. Her family, usually a source of unwavering support, now seemed to observe her with a mixture of pride and...expectation. Her mother, a former champion herself, offered congratulations with subtle suggestions for improvement. Leah understood; it wasn't criticism, but a reflection of her mother's relentless pursuit of excellence—a legacy that Leah both embraced and struggled with. Her father, less involved in the technical aspects of her riding, expressed his joy more simply. His pride was clear in the way he spoke of her achievement, his eyes shining with quiet affection. While less technical, his support was equally vital, a grounding force amidst the whirlwind of competition.

Her relationship with her older sister, Sarah, also underwent a subtle shift. Sarah, initially overshadowed by Leah's success, now seemed to view her achievement with a newfound respect. Their sibling rivalry, which had simmered beneath the surface for years, seemed to ease, giving way to a tentative camaraderie. They shared stories about riding and their lives, forging a more profound connection beyond the competitive arena. Seeing Leah's vulnerability and her hard-earned victory, Sarah recognized the sacrifices and dedication behind her success, a newfound respect that transcended their differences.

Her family did not experience the transformation alone. Her bond with Whisper deepened beyond words. Once purely functional, the quiet moments in the stable now felt sacred. Grooming Whisper became a ritual of connection, a silent conversation that transcended language. They communicated through subtle nuances; a shift in Whisper's weight, a slight twitch of her ear, a soft sigh–these became unspoken cues of understanding, a language developed through shared experiences, both triumphs and setbacks. Leah felt an overwhelming gratitude towards her equine partner, a deep appreciation for the trust and understanding underpinning their partnership.

Her relationship with Emma, however, presented a more complex evolution. The initial rivalry, fueled by competition and a clash of personalities, softened gradually. No longer perceived as solely a rival, Emma revealed a different side of herself. Their conversations, initially centered on the technical aspects of show jumping, blossomed into discussions about their lives outside the competition arena. They shared their struggles with balancing schoolwork and training, as well as their anxieties about meeting expectations and the physical and emotional demands of their sport. Relentless training and competition revealed Emma's confidence, once seen as arrogance, as resilience. Leah saw her strength reflected in Emma's and discovered a sense of mutual respect that had been absent before. The shared experience of intense competition forged a unique bond, one that recognized shared challenges and triumphs.

This newfound understanding extended beyond Emma. Leah reached out to other riders, both peers and those more experienced. She observed their riding styles, asked questions, and shared her experiences, finding a camaraderie she hadn't expected. The sense of community, once something she viewed peripherally, became a vital source of support and inspiration. She discovered a network of shared experiences, including collective challenges and triumphs, which strengthened her belief in the power of collaboration within a competitive environment. The focus shifted from individual triumph to a shared journey of excellence.

The bonds she developed extended beyond the equestrian community. Her relationship with Mr. Henderson, her trainer, moved beyond the purely professional. His guidance extended beyond the mechanics of riding, touching upon mental strategies, stress management, and the overall importance of balance in life. He became a mentor, offering insights that extended far beyond the equestrian arena. His trust in her abilities and recognition of her potential empowered her to take on new challenges in her riding and overall life. His guidance helped shape her into a stronger, more capable individual.

The changes weren't subtle; they were transformative. Leah's personal journey of growth and self-discovery transcended the competitive aspect of the competition itself, focusing instead on her development as a person. The intense pressure of the competition forced Leah to confront her weaknesses and build her resilience, sharpening her focus. Shared experiences with her horse, family, rival, and mentors shaped her character, fostering growth in empathy, understanding, and self-awareness.

She discovered the value of support systems, as well as mutual encouragement and understanding. She learned to value the community's collaborative spirit, recognizing that mutual support and encouragement could thrive even during competition. Her relationships developed beyond the competitive arena, strengthening her self-belief and enhancing her ability to navigate life's challenges with greater resilience and a sense of purpose.

The competition, once perceived as a solitary pursuit, showed her a journey shared with others — a collaborative effort that enriched her life beyond the ribbon and the trophy. The challenges overcome, the triumphs celebrated, and the relationships forged were the proper measures of her success, shaping her into a more compassionate, empathetic, and resilient young woman, ready to face future challenges with grace and determination. Her story wasn't about overcoming obstacles on the course; it was about navigating the complexities of life, one relationship at a time. And it was a journey that was beginning.

The blue ribbon showed her skill, but the deepened relationships showed her growth and development. The journey was, after all, more important than the destination. And Leah's journey, it seemed, was far from over. The next competition was on the horizon, but so were the many enriching experiences of life and the many relationships that would shape her future. Her victory at the competition wasn't a personal achievement; it was the catalyst for a profound transformation in her approach to competition and the rich tapestry of relationships surrounding her. The post-competition period was a time of profound growth and connection, reaffirming the importance of community, friendship, and family in her journey as a competitive rider and individual. The victory was a moment, but the enduring relationships forged were the legacy.

# Chapter 43: Future Aspirations

The lingering scent of sawdust and horses still clung to Leah's clothes, a comforting reminder of the adrenaline-fueled victory at the regional championships. But the echoes of the cheering crowd were fading, replaced by the quiet hum of anticipation for what lay ahead. The blue ribbon, a tangible symbol of her success, sat proudly on her dresser, a testament to her hard work and dedication. Yet, it wasn't the ribbon that held the most significance; it was the journey that led to it, the lessons learned, and the relationships forged along how truly mattered. This victory wasn't the end; it was a launching pad for further success.

Leah's future aspirations extended far beyond the immediate thrill of competition. The regional win had given her a taste of success, a confidence boost that fueled her ambition. She had tasted the pressure, felt the weight of expectation, and ultimately, emerged triumphant. This experience honed her riding skills and strengthened her mental fortitude, equipping her to face future challenges with greater resilience and determination.

Her immediate goal was simple: the national championships. The thought of competing against the country's best riders filled her with excitement and nervous anticipation. The level of competition would be significantly higher, and the pressure more intense. But Leah felt ready. The lessons she had learned, both on and off the course, had prepared her for the challenges ahead. She wouldn't be competing, but learning, growing, and pushing her boundaries even further. This wasn't about winning; it was about continuous improvement, a journey of self-discovery through the lens of equestrian sport.

Preparation for the nationals wouldn't be a solitary endeavor. Leah knew the importance of her support system. She planned to work closely with Mr. Henderson, refining her technique, focusing on specific areas for improvement identified during her recent competition. His insights extended beyond the technical aspects of riding; he understood the mental game, the importance of maintaining focus under pressure, and the significance of self-belief. Their sessions would extend beyond drills and exercises, delving into mental strategies, visualization techniques, and stress management. Mr. Henderson's guidance had been instrumental in her recent success, and she trusted his expertise to help her reach new heights.

Her relationship with Whisper would also be paramount to her success. The bond they shared was more than a rider-horse relationship; it was a partnership built on mutual respect, trust, and understanding. Their training sessions would be a collaborative dance of mutual trust and understanding. Leah planned to incorporate more cross-training activities, ensuring Whisper was physically and mentally prepared for the rigorous demands of the national competition. She would pay close attention to Whisper's physical and emotional well-being, ensuring her comfort and optimal performance. Their shared journey was a testament to the deep connection they had cultivated, a bond that would undoubtedly contribute to their success.

Her family's support would also play a crucial role in her preparation. She envisioned having more open conversations with her mother, embracing her mother's expertise, and establishing more precise boundaries. She recognized her mother's drive for excellence stemmed from love and a desire to see her daughter succeed. Her father's unwavering support, a quiet anchor in her life, would provide a comforting counterpoint to the intensity of training. Her sister, Sarah, now a source of camaraderie and shared experience, would be a valuable confidante with whom to share both triumphs and setbacks. Leah understood the importance of maintaining open communication with her family, leveraging their support to navigate the challenges ahead.

Emma, once a formidable rival, had become a respected friend and would now be a source of friendly competition and shared growth. They planned to continue post-competition conversations, discussing strategies, sharing insights, and offering mutual encouragement. The shared experience of intense competition had created an unexpected bond, one that recognized mutual struggles and triumphs. Their friendship would provide support, understanding, and a welcome distraction from the pressures of training.

Leah's aspirations extended beyond the immediate competition. She dreamed of representing her country at international equestrian events, envisioning herself riding on the world stage. This long-term goal requires dedication, perseverance, and consistent improvement. She knew this journey would involve years of hard work, continuous learning, and unwavering commitment. But the thought of representing her country, wearing the national colors, ignited a sense of pride and purpose.

Leah envisioned herself giving back to the equestrian community. She planned to mentor younger riders, sharing her knowledge and experience to help them navigate the challenges of competitive riding. She had experienced the importance of mentorship, the value of guidance and support from experienced riders. Now, she aspired to pay it forward, becoming a role model for younger equestrians. She imagined organizing workshops and clinics, sharing her expertise on technical skills, mental strategies, and the crucial importance of horse care. This aspiration reflected her passion for equestrian sport and her commitment to fostering a strong and supportive community.

Beyond the world of competitive riding, Leah also envisioned pursuing her academic interests. Leah knew that to balance her passion for equestrian sport and her educational goals, she would need careful planning and diligent time management, but she was committed to excelling in both. She knew that a strong academic foundation would provide her with diverse opportunities in the future, regardless of her equestrian career path. She was determined to excel academically, expand her knowledge further, and explore various fields of study, ensuring a well-rounded education that complemented her equestrian pursuits.

Her broader aspirations included becoming a role model, inspiring others to pursue their passions with determination and resilience. She aimed to encourage young people, demonstrating that dedication and perseverance can lead to success, even in the face of adversity. Her journey had taught her the value of hard work, the importance of self-belief, and the power of community support. She wanted to share these lessons, empowering others to embrace their challenges, cultivate their talents, and pursue their dreams with unwavering commitment. Her journey was a testament to the power of resilience, a source of inspiration for others to overcome obstacles and reach their full potential.

The future held many challenges and opportunities, but Leah embraced them all with a newfound sense of confidence and determination.

New challenges and unexpected twists would undoubtedly fill the road ahead, but Leah, fueled by her passion, resilience, and the unwavering support of those around her, was prepared to face them head-on. Her story, far from over, was unfolding, a testament to her commitment, resilience, and the transformative power of passion. The blue ribbon was a marker on a long and exciting journey filled with ambition, perseverance, and the unwavering pursuit of excellence both in the saddle and in life. The future stretched before her, a vast landscape of opportunities waiting to be explored. And Leah, with Whisper by her side and a heart full of hope, was ready to ride into it.

# Chapter 44: Continued Growth

Leah and Whisper felt the crisp fall air nip at their cheeks as they cantered across the sprawling fields bordering the stables. The national championships loomed, a looming mountain range on the horizon, but for now, the focus was on the rhythmic beat of Whisper's hooves and the steady rise and fall of her breath. The regional victory had been exhilarating — a whirlwind of adrenaline and accomplishment — but the post-competition quiet had allowed for a deeper reflection. It wasn't about the blue ribbon; it was about the subtle shifts within herself, the quiet growth that had taken root beneath the surface excitement.

This growth wasn't about improving her jumping technique or Whisper's stamina. It was about refining her mental game, a far more complex and challenging realm than any obstacle course could be. Leah realized the importance of mental preparation–visualizing each jump, anticipating challenges, and maintaining a calm, focused mind even under immense pressure. Daily mindfulness techniques, including meditation, helped her cultivate inner peace and improve concentration. She discovered that these quiet moments, spent focusing on her breath and clearing her mind, had a profound impact on her riding. She was more attuned to Whisper's subtle cues, more responsive to the nuances of the course, and ultimately, more in sync with her horse.

Mr. Henderson, her ever-wise trainer, played a crucial role in this mental refinement. He wasn't teaching her how to ride, but how to think like a champion. Their sessions moved beyond the traditional drills and exercises, delving into strategic planning and mental conditioning. He introduced her to visualization techniques, encouraging the mental rehearsal of each jump and picturing herself clearing every obstacle with ease and precision. He also stressed the importance of self-compassion, reminding her that setbacks were inevitable and that learning from mistakes was a crucial part of the journey. "It's not about being perfect, Leah," he'd often said, "it's about consistently striving to be better."

The bond with Whisper continued to deepen, developing into something more than a simple rider-horse relationship. Leah felt an intuitive connection with her mare, an unspoken understanding beyond verbal communication. Their training sessions were less about commands and more about a shared experience, a dance of mutual trust and cooperation. Leah noticed the most minor changes in Whisper's demeanor, her subtle shifts in energy and mood. She learned to adapt her training to Whisper's needs, ensuring their work was always productive, yet enjoyable for both of them. This involved incorporating more cross-training activities, like long trail rides and light work in the paddock, to keep Whisper physically and mentally engaged. She also paid careful attention to Whisper's diet, ensuring she received the proper nutrition to support her training regimen.

The support of her family became an invaluable asset. Her once-strained relationship with her mother had transformed, replaced by open, respectful conversations instead of tense exchanges. Leah realized her mother's demanding nature stemmed from a place of love. She began actively seeking her mother's advice and came to value her insightful observations. Their bond strengthened, transforming from a source of conflict into a source of encouragement and understanding. Her father's quiet support remained a constant comfort, his unwavering belief in her a reassuring presence in her life. Sarah, her sister, remained a close confidante, sharing triumphs and setbacks with a camaraderie that only siblings could understand.

Emma, once her fierce rival, now stood as a friend, a fellow traveler on the demanding path of equestrian competition. Their post-competition discussions developed into fruitful exchanges of ideas, with strategies shared and encouragement offered. They understood the unique challenges and pressures of the sport, creating a bond that transcended competition. Their shared experiences fostered a mutual respect and an appreciation for each other's strengths, creating a space for growth and shared learning. Their rivalry had become a powerful friendship, a source of inspiration and support.

The national championships arrived with a quiet intensity. Leah felt prepared, not physically, but mentally. She had honed her riding skills, strengthened her mental fortitude, and cultivated a deep connection with Whisper. The atmosphere buzzed with energy–the thrill of competition, the anticipation of success, and the weight of expectation hung heavy in the air. The pressure was intense, but Leah met it with a calm resolve, bolstered by the newfound confidence that came from months of dedicated training and self-reflection.

The competition itself was a blur of action and emotion. Leah felt a sense of focus, a clarity of purpose that had previously eluded her. She approached each jump precisely, her body attuned to Whisper's movements, her mind clear and focused. The crowd's roar faded into background noise; the only sounds that mattered were the rhythmic beat of Whisper's hooves and the whisper of the wind as they soared over each obstacle.

There were moments of doubt, fleeting uncertainties that threatened to unravel her composure. But she reminded herself of Mr. Henderson's words, the visualization techniques she had practiced, and the unwavering support of her family and friends. She breathed deeply, refocused her attention, and pressed onward.

The last jump approached—a challenging combination that demanded skill and precision. Leah felt a surge of adrenaline, but a sense of calm determination tempered her fear. She knew, in that moment, that she was ready. She and Whisper executed the jump flawlessly, a testament to their shared journey and unbreakable bond.

The judge's result confirmed what Leah knew: she had performed her best. The victory wasn't about the placing; it was about the journey, the growth, and the self-discovery that had come with striving for excellence. The blue ribbon symbolized her dedication, resilience, and unwavering belief in herself and Whisper. It was a culmination of her hard work, but more importantly, it was a launching pad for future endeavors, a testament to her endless potential.

The road ahead promised new challenges, but Leah faced them with confidence and excitement, ready to ride toward an even brighter future. The national championships were just a stepping stone, a testament to her growth in the saddle and beyond.

# Chapter 45: New Challenges

The lingering taste of victory from the National Championships was sweet, but quickly faded, replaced by a quiet hum of anticipation.

The thrill of competition had been exhilarating, a whirlwind of adrenaline and accomplishment, but the quiet moments afterwards revealed an unfamiliar landscape—one dotted with challenges yet to be faced. Leah stood at a crossroads; the path ahead was less defined than she'd imagined. The blue ribbon, symbolizing her hard work and dedication, felt both a reward and a springboard for her future endeavors. It marked a culmination, yes, but also a beginning.

The initial euphoria gave way to a subtle unease. The ease with which she'd navigated the demanding courses at Nationals had lulled her into a sense of false security. She realized that her success wasn't simply a result of her talent and hard work, but also a consequence of the specific challenges presented at that competition. The next level, she understood, would require more than refining existing skills. It required adaptation, a willingness to push her boundaries beyond what she thought possible.

This realization dawned on her during one of her quiet morning rides with Whisper. The crisp fall air, previously a source of invigorating energy, now felt charged with an additional pressure—the pressure of expectation. Once muted by the crowd's roar and the competition's intensity, the whispers of doubt now found their voice, a persistent murmur in the early morning quiet. She began questioning whether her victory had been a fluke, a lucky break in a challenging sport.

This newfound uncertainty fueled a fire within her. It wasn't the debilitating self-doubt of her past, but a healthy dose of apprehension, a recognition of the immense work ahead. She began seeking new challenges, pushing herself and Whisper beyond their comfort zones. She enrolled in advanced jumping clinics, facing more complex courses designed to test her skill and nerve. The experienced riders there, older and more seasoned, challenged her assumptions and pushed her to refine her technique, exposing her to different training methodologies and philosophies.

One such clinic introduced her to a new jumping style, focusing on a more fluid and responsive approach. Instead of relying solely on strength and precision, this technique emphasized a delicate balance of coordination and feel, a subtle dance between rider and horse. Leah initially struggled, her ingrained habits resisting the subtle adjustments required. Her body, accustomed to a more forceful style, found the transition awkward and unfamiliar. However, she persevered, driven to broaden her skill set and expand her horizons.

Whisper faced new challenges as well. The higher-level training demanded more physically and mentally. Under Mr. Henderson's guidance, Leah introduced a rigorous program that included hill work, interval training, and specialized exercises to boost Whisper's agility and stamina. They also focused on precision maneuvers, refining their ability to execute complex movements with accuracy and grace.

The bond between Leah and Whisper deepened through these trials. The shared struggles forged a connection that extended beyond the confines of the riding arena. Leah learned to interpret Whisper's subtle cues with greater sensitivity, her body becoming attuned to the mare's energy shift. She recognized that a successful partnership required skillful riding and a deep understanding of her horse's physical and emotional needs. This new level of communication transcended verbal commands, becoming a silent dialogue woven through their shared training.

Beyond the physical challenges, Leah encountered new social dynamics. The intensity of the competitive equestrian world heightened. The camaraderie she'd found with Emma continued to blossom, but the pressure of aiming for international competitions introduced new rivalries, creating both healthy competition and subtle anxieties. She learned to navigate the complexities of the equestrian social scene, understanding the delicate balance between ambition and collaboration.

There were moments of frustration, of setbacks that threatened to derail her progress. There were falls, misjudged jumps, and moments of discouragement.

But Leah learned to embrace these setbacks as valuable lessons, each stumble a stepping stone on the path toward improvement. She developed a resilience that transcended the riding arena, enabling her to face personal challenges with the same strength and determination.

She also discovered the value of mentorship. Seeking advice from seasoned professionals, observing their training methods, and understanding their approaches expanded her knowledge base and provided her with valuable perspectives on the intricacies of high-level competition. Unlike her sometimes-overbearing mother, these mentors offered constructive criticism and encouragement, helping her refine her skills without sacrificing her confidence.

The changes extended beyond the riding arena and into her personal life. The newfound confidence she had gained had a positive impact on other aspects of her life. She became more assertive in her academic pursuits, demonstrating a similar level of dedication and perseverance she had applied to her equestrian endeavors. Her relationships with her family and friends also deepened, and her newfound strength and self-awareness made her more capable of fostering healthier and meaningful connections.

The new challenges, far from discouraging, fueled Leah's passion and reinforced her determination. They presented an opportunity to redefine her goals and set her sights on even higher aspirations. The national championships had been a milestone, but they weren't the ultimate destination. Leah knew her journey as an equestrian was a continuous growth process, a lifelong pursuit of excellence. Uncertain as the road ahead was, she eagerly embraced the new challenges and adventures awaiting her.

Looming large were the upcoming international competitions, a significant step forward in her career. Immense pressure, fierce competition, and even higher stakes would be present. But Leah felt prepared, not merely in terms of her riding skills, but also her mental resilience and her unwavering belief in herself and Whisper. She had learned to navigate the complexities of the equestrian world, both on and off the course.

# Chapter 46: Repairing Relationships

The quiet hum of anticipation preceding the international competitions wasn't solely about the jumps and the ribbons. It was also about people–the people who had supported her, challenged her, and, sometimes, let her down. Leah reflected on the relationships that had shaped her journey, relationships that needed mending before she could fully embrace the next phase of her competitive career.

First on her list was her mother. Their relationship was complex, a delicate dance between fierce ambition and unspoken resentment. As a former champion herself, Leah's mother projected her unfulfilled dreams onto Leah, creating a pressure cooker environment where even minor mistakes seemed catastrophic. The constant comparisons, the subtle criticisms, and the overwhelming expectations had created a chasm between them, a silence punctuated only by tense conversations and unspoken resentments.

Leah approached the issue directly. She sought her mother, not in the stables' bustling atmosphere, but in the quiet solitude of their family home. They sat in the sun-drenched conservatory, the scent of blooming jasmine filling the air, starkly contrasting with the usual tension that permeated their interactions.

"Mom," Leah began, her voice surprisingly steady, "I need to talk to you."

Her mother looked up, her eyes betraying a flicker of apprehension, a mixture of anticipation and trepidation. Leah continued, carefully choosing her words, "I know things haven't been easy between us. I know I haven't always met your expectations, and maybe I haven't always understood why you pushed me so hard."

She paused, giving her mother a chance to respond. The silence stretched, heavy with unspoken years. Finally, her mother spoke, her voice laced with a hint of vulnerability, "Leah, I... I wanted you to have everything I never had. I wanted you to succeed where I failed."

Tears welled in her mother's eyes. Leah reached across the small table separating them, placing her hand over her mother's. Years of emotional distance vanished with physical touch. "I get it," Leah murmured, "However, your expressions of affection... sometimes cause me pain."

The conversation flowed from there, a torrent of long-held emotions, unspoken regrets, and finally voiced regrets. Leah expressed her frustration, anger, and disappointment, but did it with empathy, acknowledging her mother's intentions, even if the methods were flawed. Her mother truly listened, offering apologies and expressing her remorse.

The conversation didn't erase years of tension, but laid the foundation for a healthier, more authentic relationship. It was a testament to Leah's maturity, her ability to see beyond the hurt and understand the complexities of her mother's motivations.

Another relationship that required attention was her friendship with Emma. Their rivalry had been intense, a fiery clash of wills that sometimes overshadowed their mutual respect and admiration for each other's skills. The competition had pushed them both, introducing moments of friction, veiled insults, and unspoken jealousy. Their shared ambition sometimes felt like a battlefield where even minor gestures carried significant weight.

Leah realized their competitiveness had obscured a genuine connection, a shared passion for the sport. She sought Emma, proposing a joint training session, not as rivals, but as equals. They spent hours together, sharing their techniques, offering constructive criticism, and pushing each other to improve. The experience strengthened their bond, transforming their rivalry into a supportive partnership.

Their shared passion for equestrian sports became a bridge, transcending the superficial aspects of competition and revealing their deep-seated respect and admiration for each other. They discovered they were more alike than they had ever realized. Their shared dedication to their sport forging a strong and lasting friendship.

Beyond her close relationships, Leah also made amends with some of her fellow competitors. There were instances of harsh words, misinterpreted gestures, and subtle acts of sabotage that had overshadowed their shared passion for the sport. She approached them with an apology, taking responsibility for her actions and acknowledging the negativity that had clouded their interactions. This reconciliation showed her maturity and ability to see beyond petty rivalries. It also showed growth in her emotional intelligence, an understanding that success in any endeavor was not simply about individual achievements, but about building strong, supportive relationships with those who shared a similar passion.

The process of repairing these relationships wasn't easy. It required humility, empathy, and a willingness to confront her flaws. It meant acknowledging the hurt she had caused because of her mistakes and actively seeking reconciliation. But the effort was worth it. The relationships were stronger, more authentic, and provided a more profound sense of belonging and support. The mending of these relationships extended beyond mere forgiveness. It brought about a significant shift in Leah's perspective, teaching her the importance of empathy and understanding in all areas of her life. The process enhanced her maturity, reinforcing her emotional intelligence and profoundly shaping her character.

As the international competitions drew near, Leah felt a profound sense of peace from her physical preparation and the emotional clarity she had gained. She felt ready to compete as a talented rider and a well-rounded individual, equipped with the emotional resilience and strong relationships necessary to navigate the challenges ahead. The road ahead was still long and arduous, filled with potential setbacks, but Leah knew she was ready, fortified by her personal and equestrian achievements. The accurate measure of her success wasn't about winning; it was about growing as a person, repairing bridges, and nurturing the meaningful connections that enriched her life both on and off the course. She understood excelling required not only show jumping mastery but also skillful personal relationship management. The quiet confidence she exuded showed how strong the bridges she'd carefully built were. A new phase of her life, both riding and otherwise, was on the horizon.

# Chapter 47: Forging New Bonds

The quiet confidence within Leah wasn't solely a product of her improved riding technique or the meticulous preparation for the upcoming international competition. It stemmed from a deeper, more profound source: the forging of new bonds, the expansion of her support network beyond the familiar faces of her family and long-time rivals. This wasn't a sudden shift, but a gradual process, a carefully constructed tapestry woven from threads of empathy, understanding, and shared passion.

One of the unexpected connections Leah forged was with Mr. Henderson, the gruff but ultimately kind-hearted groundskeeper at the stables. Initially, she had seen him only as a man who quietly tended to the grounds, a figure whose presence was as constant as the scent of freshly cut hay. But as she spent more time at the stables, practicing late into the evening, his quiet strength and unwavering dedication drew her.

One evening, after a grueling training session, Leah found Mr. Henderson meticulously repairing a damaged fence. The setting sun cast long shadows across the paddock, painting the scene in hues of orange and purple. Hearing the day's exhaustion weighing down on her, Leah approached him. The conversation that followed was initially tentative, a hesitant exchange of pleasantries. But as they talked, Leah discovered that Mr. Henderson, despite his taciturn nature, possessed a wealth of knowledge about horses, their quirks, and their needs. He shared stories about his experiences with horses, anecdotes that painted a vivid picture of a life deeply intertwined with the animals he cared for.

These conversations became regular, small moments of connection, transcending the usual trainer-rider dynamic. Mr. Henderson offered Leah insights into horse psychology, teaching her to read subtle animal behavior cues and understand their anxieties and strengths. He became an unexpected mentor, and his practical wisdom and deep-rooted experience provided a counterpoint to the more formal training she received from her coach. His quiet support became a steady anchor, grounding Leah and helping her to approach her, riding with a renewed sense of perspective.

Leah also cultivated a connection with a group of younger riders taking part in the junior competition alongside her. Initially, she had seen them as mere rivals, competitors vying for the same ribbons and accolades. However, as the competition approached, their enthusiasm, unwavering belief in themselves, and shared a love for the sport drew her in. She began mentoring them, sharing her expertise and experience. This generosity strengthened her understanding of the sport and broadened her emotional horizons. She found satisfaction in helping others grow, nurturing their passions, and witnessing their triumphs.

The young riders offered Leah a different perspective on the pressures of competition. Her life of intense scrutiny and high expectations contrasted with their uninhibited joy and unburdened approach. Their unadulterated passion for the sport reminded her of the simple joy of riding, the intrinsic satisfaction of connecting with an animal, and working together towards a common goal. Their friendship provided a vital emotional buffer, shielding them from the blows of self-doubt and the anxieties of competition.

In the lead-up to the competition, Leah's expanded network offered invaluable emotional support. Mr. Henderson's calming presence and practical advice eased her pre-competition jitters, while the unwavering enthusiasm of the younger riders reminded her of the simple joy of riding. Even her newfound friendship with Emma, now cemented by mutual respect and shared training sessions, offered a sense of camaraderie and shared experience that diminished the solitary weight of competition.

The bonds Leah had forged extended beyond the realm of equestrian sports. She reconnected with an old friend from her childhood, a fellow book enthusiast who shared her passion for literature and storytelling. This unexpected connection offered a welcome escape from the intense world of competitive riding, providing a space for laughter, conversation, and intellectual stimulation. These moments of respite were critically important.

The friendships Leah nurtured weren't solely about receiving support; they involved active participation, offering help and encouragement to others. She volunteered at a local riding school, where she worked with children starting their equestrian journey. This experience deepened her understanding of teaching, patience, and empathy. The children's innocent enthusiasm and eagerness to learn were infectious, reminding her of the pure joy that lay at the heart of the sport.

This broadened sense of community strengthened Leah's resilience and self-belief. She realized success wasn't about winning individual accolades and building strong, supportive relationships within and beyond the equestrian world. The friendships she cultivated weren't peripheral connections; they became integral parts of her support system, helping her navigate the challenges of competition, celebrate her victories, and learn from her setbacks.

As the international competition approached, Leah approached it with a newfound confidence in her riding skills and her capacity to build and sustain meaningful relationships. This confidence extended beyond the riding arena, permeating every aspect of her life, from her interactions with her family to her connections with fellow competitors and her newly forged friendships. Her painstakingly constructed support network provided a vigorous defense against self-doubt and intense competition.

The number of ribbons she collected or the titles she won would solely define the accurate measure of her success. Her profound connections, strong friendships, and significant influence on others' lives all served as measures of her depth. She realized that achieving excellence in show jumping required not just skill, but also empathy, understanding, and building relationships that nurtured her spirit and enriched her life. New connections strengthened her, preparing her for future difficulties and changing her deeply and permanently.

# Chapter 48: Embracing Teamwork

The quiet hum of anticipation vibrated through the stables the day before the Grand Prix. The air crackled with nervous energy, a palpable tension that hung heavy like the scent of freshly oiled leather and horse sweat. Leah, usually consumed by her anxieties, found herself oddly calm. This wasn't the result of some newfound magical confidence, but a quiet understanding — a deep-seated acknowledgment of the intricate network of support woven around her. She was no longer a rider preparing for a competition; she was a part of a team, a vital piece in a complex puzzle of shared ambition and collective effort.

This realization struck her during a late-night training session with her coach, Mrs. Davies. Mrs. Davies, renowned for her meticulous attention to detail and unwavering standards, pointed out a subtle flaw in Leah's approach to a particular jump. Instead of reacting defensively, as she might have done in the past, Leah listened intently, absorbing the criticism with a newfound receptiveness. She realized that Mrs. Davies's feedback wasn't a personal attack but a collaborative effort towards improvement. It was a testament to the trust and mutual respect that had grown between them over months of intense training.

Beyond her coach, Leah considered the contributions of her groom, Liam. He was more than someone who cleaned tack and cared for her horse, Apollo. He was a silent observer, with a keen eye for even the slightest change in Apollo's behavior, a constant source of reassurance and practical support. Liam's quiet dedication had allowed Leah to focus on her riding, knowing that Apollo was in capable hands, his every need meticulously attended to. He'd noticed the slight lameness in Apollo's left foreleg days before anyone else, leading to early intervention and preventing a potential catastrophe. Leah had previously taken this contribution for granted but now recognized it as crucial to her success.

Then there was the farrier, Mr. Fitzwilliam's expertise perfectly balanced and protected by Apollo's hooves. His knowledge of equine podiatry was unmatched, and his meticulous attention to detail had played a significant role in Apollo's performance. The subtle adjustments he made to Apollo's shoes, invisible to the untrained eye, had made a tangible difference in the horse's comfort and agility.

The veterinarians also played a crucial role. Their regular check-ups and prompt attention to minor ailments ensured Apollo remained healthy and fit. They offered expertise beyond simply treating illness; Leah had witnessed firsthand how a slight imbalance in Apollo's diet could affect his energy levels and jumping technique. The veterinary team had provided the essential expertise to navigate these potential issues, proactively maintaining Apollo's well-being.

Even the most minor details, previously overlooked, now took on new significance. With meticulous care, the stable staff ensured a clean and safe environment for the horses by cleaning and maintaining the stables. The officials who organized the competition ensured fair play and a smooth running of the event. The spectators, whose cheers and support fueled the riders' energy. Each individual played a vital part in the intricate dance of competitive equestrian sports.

Leah realized that her performance wasn't solely dependent on her skills. It was a product of the collective effort of many individuals, each contributing their expertise and support to the common goal. This realization instilled in her a profound sense of gratitude and a newfound appreciation for the collaborative nature of the sport.

Seeking opportunities, she expressed her appreciation. Liam's meticulous care of Apollo and keen observation skills earned him profuse thanks from her, along with an acknowledgement of his dedication. She thanked Mr. Fitzwilliam for his expertise, explaining how much Apollo's comfort affected his performance. She even made a point of thanking the stable staff for maintaining such a pristine environment, emphasizing the impact of a clean and organized space on horse and rider well-being.

This newfound appreciation extended beyond the immediate team surrounding her. Leah actively sought to collaborate with fellow competitors. She offered Emma a helping hand during a challenging training session, sharing her experience with the specific obstacles that Emma was facing. She assisted some of the younger riders, sharing techniques and offering encouragement. This wasn't an act of altruism alone; it was a recognition of the reciprocal nature of support. By assisting others, Leah deepened her understanding of the sport and strengthened her bonds with fellow competitors. The camaraderie in these interactions enriched her experience, transcending the pressures of competition, and fostered a deeper appreciation for the shared passion that unites those who love equestrian sports. These collaborations weren't about improving individual performance, but creating a supportive and collaborative environment where everyone could thrive.

The Grand Prix dawned on a bright and clear day, with the sun illuminating the meticulously prepared arena, setting the stage for a spectacular event. While meticulously preparing Apollo for the upcoming competition, Leah unexpectedly discovered a profound sense of tranquility and serenity that transcended any emotion she had previously encountered. Although the pressure remained, the shared responsibility and mutual support lightened considerably the burden of the team members. She had the support of her friends and family. She found herself surrounded by a group of people that supported her in the Grand Prix and her future.

The crowd's roar washed over her as she rode into the arena, but it no longer felt overwhelming. It was a wave of collective energy, a testament to the power of shared passion and collaborative effort. She rode with a renewed confidence, not in her abilities, but in the team's strength that stood beside her. The result of the competition, she knew, would reflect not her skills but the collective effort, the shared dedication, and the collaborative spirit that had brought her to this moment.

# Chapter 49: Seeking Mentorship

The quiet confidence that settled over Leah the day before the Grand Prix didn't vanish with the sunrise. Instead, it developed, morphing into a focused determination. The realization that her success wasn't solely reliant on her skills had sparked something within her–a hunger for more profound knowledge, a desire to learn from those who had already traversed the challenging path she was now navigating. This realization led her to seek mentorship, a conscious decision to expand her circle of support beyond her immediate team.

Her first approach was to the legendary Isabelle Dubois, a retired show jumper whose name echoed through the annals of equestrian history. Isabelle's impeccable technique, unwavering composure under pressure, sharp intellect, and insightful coaching style were renowned. Leah, summoning all her courage, approached Isabelle after a practice session, her heart hammering a frantic rhythm against her ribs.

"Ms. Dubois," Leah began, her voice slightly trembling. I've admired your career for as long as I can remember. Would you be open to mentoring me?"

Isabelle, a woman whose face held the etched lines of countless competitions and whose eyes held the wisdom of years spent at the pinnacle of the sport, studied Leah with a keen, assessing gaze. Then a slow smile played on her lips. "Mentorship is a two-way street, young lady," she replied, her voice a low, melodious murmur. "It requires dedication, commitment, and a willingness to learn. Are you prepared for that?"

Leah's nod was firm. "Yes, ma'am. I'm ready to work hard."

Isabelle agreed to meet with Leah once a week, offering her insights into strategic course planning, mental preparation, and the subtle nuances of horse-rider communication. Their sessions weren't merely technical discussions; they were in-depth explorations of the psychological aspects of competition, examining Isabelle's strategies for managing pressure, channeling anxiety, and maintaining focus under intense scrutiny. Isabelle helped Leah develop visualization techniques, teaching her to rehearse each jump mentally, visualize flawless execution, and build self-belief to reduce self-doubt's impact.

Beyond Isabelle's technical expertise, Leah found immeasurable value in Isabelle's wisdom on navigating the complexities of the equestrian world. Isabelle's insights extended beyond the arena; she shared anecdotes from her illustrious career, offering perspectives on handling sponsorships, dealing with media pressure, and maintaining professionalism within the often-competitive environment of professional equestrian sports. She emphasized the importance of preserving integrity and fair play, stressing the importance of respecting fellow competitors and their horses.

Simultaneously, Leah sought guidance from another source: a renowned equine veterinarian, Dr. Ramirez. While her veterinary team had provided excellent care for Apollo, Leah recognized the value of consulting a specialist who could offer a broader perspective on equine health and performance. Dr. Ramirez, a passionate advocate for preventative care and equine well-being, agreed to mentor Leah, offering insights into the subtle indicators of equine health that even experienced grooms might miss. Their conversations revolved around nutrition, exercise physiology, and the preventative measures Leah could take to optimize Apollo's health and performance.

Dr. Ramirez emphasized the importance of understanding Apollo's needs, stressing that what works perfectly for one horse might harm another. They delved into the intricate relationship between diet, training, and overall well-being, emphasizing that peak performance hinges on a holistic approach to equine care. Dr. Ramirez trained Leah to spot when Apollo seemed stressed, tired, or uncomfortable, so Leah could prevent minor problems from becoming big ones. This mentorship went beyond simply providing veterinary advice. It cultivated a deeper understanding of the crucial partnership between horse and rider.

The combined mentorship from Isabelle and Dr. Ramirez had a profound impact on Leah. She aimed to improve her riding, enhance Apollo's condition, and expand her overall understanding of the sport. She learned to appreciate the interwoven elements that contribute to success—the technical skills, mental fortitude, and unwavering commitment to both personal and equine well-being. The experiences strengthened her confidence and instilled in her a more profound sense of responsibility, not only for her performance, but also for the welfare of her horse.

Leah's newfound approach extended beyond formal mentorship. She actively sought conversations with other experienced riders, observing their training techniques, listening to their strategies, and gleaning insights from their experiences. She learned from successes and setbacks, recognizing that even the most accomplished riders faced challenges and setbacks throughout their careers.

This proactive approach to learning transformed Leah's relationship with her fellow competitors. Instead of viewing them as rivals, she saw them as valuable sources of knowledge and inspiration. The collaborative spirit she'd discovered before the Grand Prix strengthened. Through lively discussions about training techniques, coarse strategies, and mental preparation, she not only fostered mutual respect and collaboration among her fellow competitors, but also engaged in a spirit of shared learning and camaraderie.

The mentorship wasn't a onetime event, but an ongoing process — a continuous journey of learning and growth. Each conversation, each training session, and every piece of advice contributed to the evolution of Leah's riding style, her mental approach to competition, and her overall understanding of the sport. The support, guidance, and collaboration helped her become a more confident, well-rounded, and capable equestrian athlete, preparing her for the upcoming Grand Prix and the challenges ahead in her equestrian career. And as she prepared for her next competition, the lessons learned, and the bridges built would serve as a sturdy foundation for the future. The quiet hum of the stables now held a different resonance—a calm symphony of shared experience, mutual respect, and the enduring power of mentorship.

# Chapter 50: Looking Ahead

The cool evening air carried the scent of freshly turned earth and hay, a familiar fragrance that always calmed Leah. She stood by the stable, Apollo's warm breath puffing gently against her hand. The Grand Prix was over, a whirlwind of adrenaline, anxiety, and ultimately, triumph.

The mentorship from Isabelle Dubois and Dr. Ramirez had been transformative, offering her insights far beyond the technical aspects of show jumping. Isabelle had instilled in her advanced riding techniques and a deep understanding of the mental game, as well as a strategy for managing pressure and harnessing nerves to achieve peak performance. Their sessions had gone beyond the arena. Isabelle had shared stories of resilience, of setbacks overcome, such as the unwavering commitment required to reach the highest levels of competition. She had spoken about the importance of grace under pressure, not only in the ring but also in the face of criticism or setbacks. Isabelle's wisdom extended beyond the sport, offering invaluable advice on navigating sponsorships, managing media scrutiny, and upholding ethical conduct. She had stressed the importance of treating every competitor with respect, regardless of skill level, and always acknowledging the significance of the horse-rider partnership.

Dr. Ramirez's mentorship had an equally profound impact. Their conversations had moved beyond immediate veterinary care; they had delved into the holistic well-being of Apollo, exploring the intricate connections between diet, training, and mental health in horses. Dr. Ramirez's expertise extended far beyond the basics; she taught Leah how to read Apollo's subtle cues, recognizing the signs of stress or discomfort, and proactively address them before they became serious problems. Leah had learned to expect Apollo's needs, adjusting training regimes based on his responses, and ensuring he was always thriving, not surviving. It was more than care; it was an acknowledgment of the deep partnership between horse and rider, a relationship of trust and mutual respect that underpins all aspects of competitive equestrian sports.

But the students hadn't confined their learning to these formal mentorships.

Leah proactively sought knowledge from every source available. Engaging in conversation with her fellow riders, she keenly observed their riding techniques, posed insightful questions, and listened intently as they shared their inspiring tales of both triumph and setbacks. Through her experience, she came to understand and value the diverse methods used in training, realizing that there wasn't just one correct approach, but an array of techniques that proved effective depending on the unique needs and characteristics of each horse-rider pair. Observing the groomsmen and stable hands, she absorbed their extensive experience, practical skills, and committed care for the horses. Her research showed that equestrian knowledge came from many sources.

Her relationship with her fellow competitors underwent a significant transformation as well. Initially viewed as rivals, they became valuable sources of inspiration and support. The shared challenges and mutual respect for the skill and dedication required in the sport created a bond that transcended competition. She was now engaged in constructive dialogue with Emma, their initial rivalry softened by a mutual appreciation for their shared passion. They discussed course strategies, training techniques, and the mental hurdles they faced.

This collaborative spirit extended beyond the arena; they often shared a meal or unwound together after a challenging day, discussing the equestrian aspects and other aspects of their lives. This shared experience strengthened their bonds and helped Leah to see the importance of community and support within a seemingly competitive world.

The future stretched before Leah, not as a daunting expanse of uncertainty, but as an exciting landscape of possibilities. The Grand Prix victory was a significant milestone and a launching point for future adventures. She envisioned competing in increasingly prestigious events, pushing her limits, and furthering her knowledge. She saw herself mentoring younger riders, sharing her lessons, guiding them through the challenges and triumphs of competitive equestrian sports. She envisioned continuing her work with Dr. Ramirez, deepening her understanding of equine well-being, and contributing to the advancement of equine care and preventive medicine. She pictured herself creating her own training program that used her insights and what she'd learned from her mentors.

Her vision extended beyond the competitions and accolades. She championed responsible horse ownership, advocating for equine welfare and ethical equestrian practices. She understood how her success as a rider could promote responsible practices, advocate for change, and build a more sustainable and ethical future for the sport. Her vision was to contribute to equine research, leveraging her skills and resources for the horses she cherished.

The change wasn't in her riding, but in her perspective. Leah had found her voice, not only as an equestrian athlete but also as a leader and advocate for the sport. She understood the importance of community, the power of collaboration, and the enduring significance of mentorship. She had learned to embrace her vulnerability, acknowledge her strengths and weaknesses, and approach every challenge with renewed courage and determination.

As she looked out across the star-dusted sky, the quiet hum of the stables was a comforting lullaby, and Leah felt a profound sense of gratitude. The journey hadn't been easy, but it had been worth it. The bridges she had built—with her mentors, her fellow competitors, and, above all, with Apollo—were strong and enduring. The future held immense potential, and she was ready to compete, contribute, inspire, and leave her indelible mark on equestrian sports. Although the road ahead was long, uncertain, and exciting, Leah felt completely prepared for anything.

# Chapter 51: Setting New Goals

The quiet hum of the stables faded as the sun began its ascent, painting the eastern sky in hues of apricot and rose. Still basking in the afterglow of her Grand Prix victory, Leah felt a different exhilaration this morning–not the adrenaline rush of competition, but the calm confidence of a goal achieved and new horizons beckoning. The victory wasn't merely a culmination but a springboard, propelling her toward a future brimming with possibilities. She knew, with a certainty that resonated deep within her soul, that this was the beginning.

The previous weeks had been a whirlwind of activity. She'd received congratulatory messages from sponsors, media requests for interviews, and invitations to various equestrian events. The attention was flattering, yet it didn't distract her from the inner reflection that the competition had sparked. Focusing on both technical skills and mental fortitude, she'd spent hours analyzing her performance. She scrutinized each jump, movement, and fleeting emotion. She identified areas where she excelled and could improve–not in her riding technique, but in her approach to the sport.

This introspection led her to articulate a set of new goals, goals that extended far beyond winning ribbons and accolades. They were ambitious, some might even say audacious, but they were her own, deeply personal aspirations born from the lessons she'd learned and the experiences she'd lived. Her first goal was to hone her riding skills further. She planned to dedicate herself to consistent, rigorous training, seeking advanced clinics and workshops with renowned international riders. She wanted to master new techniques, refine her style, and develop a stronger connection with Apollo. This wasn't about physical prowess; it was about understanding the subtle nuances of communication between horse and rider, achieving a level of harmony that transcended words. She envisioned herself riding with an effortless grace, a seamless synchronization between herself and her horse, an almost telepathic understanding.

Dr. Ramirez's mentorship fostered her enhanced understanding of equine well-being, which directly tied to her second goal. Ramirez. She wanted to help riders take better care of their horses and share her knowledge with others. This involved conducting research, attending conferences, and actively participating in discussions related to equine health, nutrition, and ethical treatment. She envisioned herself delivering workshops on equine welfare, sharing her knowledge and expertise with other riders, trainers, and horse owners, contributing to a broader movement aimed at enhancing the lives of equine animals.

Her third goal was to give back to the sport that had given her so much. To mentor aspiring young riders, she wanted to share her technical skills and experiences, helping them overcome self-doubt, build resilience, and understand perseverance. Guiding them through challenges and joys, she envisioned helping them build confidence, overcome obstacles, and develop unique riding styles. She saw herself as a beacon of support and inspiration, sharing the hard-won wisdom she had accumulated along the way. She aimed to establish a supportive network for inexperienced riders, promoting collaboration and a sense of community.

Her fourth goal, perhaps the most ambitious of all, was to establish her equestrian training center. She envisioned a facility dedicated to promoting holistic horse care, providing high-quality training, and nurturing the next generation of equestrian athletes. She saw it as a place where the welfare of the horses would be paramount, where ethical practices were not merely words but a cornerstone of the program. The center would be a testament to her values, a reflection of everything she had learned and the principles she embraced.

But these goals weren't abstract ideals but carefully constructed steps on a well-defined path. Leah developed a detailed plan, outlining specific timelines, measurable objectives, and resources required to achieve each goal. She researched potential sponsorships, explored funding avenues, and identified mentors and collaborators who could offer guidance and support. She was strategic and realistic, yet never lost sight of her ultimate vision.

Outside of work, her commitment to her new goals was clear of everything she did, affecting her personal life and relationships with intense passion. To improve her physical and mental health, she redesigned her daily routine, adding focused exercise and mindfulness to increase her fitness. Prioritizing both physical and psychological well-being for peak performance, she focused on rest and recuperation. Although she pushed herself and Apollo to achieve more athletically during training, she prioritized their comfort and well-being. Their bond deepened over time, transcending their initial rivalry. Built in mutual respect, the partnership blossomed into a journey of shared growth and exciting discoveries.

Besides her other commitments, she cultivated strong bonds with both Isabelle Dubois and Dr. Ramirez, relying on their ongoing mentorship and help for continued success. Their expertise and wisdom meant a great deal to her, but even more so did their unshakeable faith in her capabilities and future success. She came to understand that effective mentorship is not a singular act, but an enduring process, characterized by a consistent and reciprocal flow of knowledge and mutual support. To ensure collaboration and consistent progress, she established a routine of regular meetings with both individuals, diligently providing updates on her advancements, actively soliciting their feedback, and consistently expressing a genuine appreciation for their valuable perspectives.

She nurtured her relationship with her fellow competitors, particularly Emma. Their earlier rivalry had transformed into mutual respect and understanding forged in the crucible of competition. They continued to share training tips, offer support, and celebrate each other's successes. This newfound camaraderie extended beyond the arena to shared meals, casual conversations, and an understanding of the shared dedication and passion for the sport.

As Leah embarked on this new chapter of her life, she felt a quiet determination. No longer: She was not a young rider striving for victory; Her goals were not personal aspirations. The journey ahead would be long and challenging, but Leah was ready. She possessed the skills, knowledge, a supportive network, and, most importantly, an unwavering belief in her potential. The quiet hum of the stables now resonated with the promise of a brighter future shaped by her ambition, dedication, and profound love for the sport.

As the sun ascended in the sky, its golden beams cast a glow of approval over her resolute expression, lighting up her features as she readied herself to confront the challenges of the day that lay ahead, a day filled with boundless opportunities. With great anticipation, the global equestrian sports community eagerly awaited Leah, who had meticulously prepared herself to confront the myriad challenges and embrace the exciting opportunities that awaited her on the grand stage of international competition.

# Chapter 52: Embracing the Future

The crisp fall air nipped at Leah's cheeks as she led Apollo into the paddock, the rising sun painting the dew-kissed grass with gold. The victory at the Grand Prix felt less like a distant memory and more like the foundation upon which she was building her future. It wasn't about winning; it was about the journey, the lessons learned, and the growth that came with it. The quiet confidence within her was palpable, a tangible sense of purpose resonating with every stride of Apollo's powerful legs.

This new chapter wasn't about refining her riding skills and redefining her relationship with the sport. She saw it not as a relentless competition but as a collaborative partnership with her horse, her trainers, fellow competitors, and even the horses themselves. She felt a deep sense of responsibility and stewardship toward the well-being of her equine partners. This wasn't simply about winning blue ribbons, but about fostering a profound respect and understanding between humans and horses.

Her ambitions and plans reached far beyond the scope of her own personal goals and aspirations, encompassing a much wider vision. In her vision of the future, she saw an equestrian world transformed into a more inclusive, sustainable, and ethically sound community, where all are welcome and the well-being of horses and the environment is paramount. Recognizing the urgent need to elevate the standards of responsible horse care and actively promote the well-being of these majestic creatures, she dedicated herself to this cause. She wanted to use her platform, her newfound influence, to advocate for changes within the sport and create a more fair and compassionate environment for both horse and rider.

Early in her development, her trainers emphasized upgrading her skills. To improve, she meticulously reviewed her training, identified weaknesses, and focused on refining her technique. Her aspirations extended beyond mere horsemanship; This involved pushing herself and Apollo to their physical and mental limits, exceeding their comfort zones while always prioritizing careful conditioning and injury prevention strategies. With Isabelle Dubois's close guidance, she developed a personalized training plan balancing intense workouts with proper rest for peak performance. Isabelle's sharp eye caught a subtle shift in Leah's demeanor; she was more self-aware, closer to Apollo, and utterly dedicated to success.

Leah's commitment extended to the mental aspect of the sport. She sought the guidance of sports psychologist Dr. Chen to help her develop coping mechanisms for the inevitable pressures of competition. She learned techniques to manage stress, control her anxieties, and maintain her focus under pressure. Dr. Chen helped her reframe her thinking, replacing self-doubt with self-belief, and view challenges as both obstacles and opportunities for growth. This focus on mental fortitude proved invaluable, allowing her to approach competitions with a calm confidence that was truly inspiring.

The collaboration with Dr. Ramirez on equine welfare significantly enhanced her existing knowledge and expertise in this important area of animal health. Immersed in her research, she dedicated countless hours to pouring research papers, actively participating in informative seminars, and engaging in collaborative discussions with leading experts within her field of study. With a firm commitment to improving equine welfare, she set her sights on becoming a leading voice within the equestrian community, educating others about responsible horse care, advocating for implementing ethical practices, and fighting for the establishment of higher standards. She pictured a future where the horse's well-being was of utmost importance, a future where its welfare was always the top priority.

The idea of establishing her equestrian center was no longer a distant dream, but a concrete goal. She began drafting a business plan, researching potential locations, and exploring avenues for funding. She envisioned a facility that prioritized the welfare of the horses and the development of young riders. They would prioritize ethical practices, treat horses with respect and dignity, and nurture and support young riders. She sought advice from successful entrepreneurs in the equestrian world, learning from their experiences and building upon their wisdom.

Leah also reached out to fellow competitors, not as rivals, but as colleagues and friends. She shared her insights and expertise with them, offering support and encouragement. The fierce competition of the past had transformed into a collaborative spirit. She realized that the equestrian community was stronger and could achieve more by working together towards shared goals. She mentored younger riders, sharing the hard-won lessons she had learned about resilience, self-belief, and perseverance.

The transition wasn't without its challenges. There were moments of self-doubt, moments when her goals seemed overwhelming. But Leah had learned to embrace these moments, to see them not as setbacks, but as opportunities for learning and growth. She drew strength from her support network—Isabelle, Dr. Ramirez, Dr. Chen, her family, and her fellow riders. They offered guidance, encouragement, and unwavering belief in her potential.

Leah's progress toward her goals continued steadily as the weeks bled into months, marking a significant period of advancement. To prepare for opening her equestrian center, she successfully got sponsorships, acquired the funding, and located a perfect site for its construction. With a focus on equine welfare and safety, she embarked on the design of the facility, implementing the highest standards at every stage of the process to guarantee a safe and comfortable environment for the animals. Working alongside a team of architects, engineers, and various other professionals, she meticulously crafted a space designed to be both highly functional and profoundly inspiring.

The process was demanding, requiring long hours, dedication, and a relentless commitment to excellence. But Leah found herself energized by the challenge. She had discovered a passion for leadership, a deep sense of purpose that went beyond personal achievement. She saw herself as a steward of the sport, nurturing its growth while preserving its values and ethics. As she looked out over the rolling hills surrounding her future equestrian center, a sense of profound gratitude washed over her. It hadn't been easy, but she had overcome her self-doubt, harnessed her ambition, and transformed her passion into a tangible reality.

She knew the journey ahead would be long and challenging, but she was ready. She had the skills, vision, support, and, most importantly, the unwavering belief in herself and her ability to positively affect the world of equestrian sports. The sun set, casting a warm golden glow over the land, as if approving of the new chapter unfolding before her. This was the beginning. Leah, ready to ride towards it, saw a bright future. The journey was far from over, but with Apollo by her side, Leah was prepared to embrace every challenge, opportunity, and new beginning. The world of equestrian sports was waiting, and Leah was ready to meet it head-on with her unwavering spirit and commitment.

# Chapter 53: Personal Growth

The scent of freshly turned earth and pine needles filled the air as Leah sat on the porch of her family's old farmhouse, a steaming chamomile tea warming her hands. The sprawling fields, usually a vibrant green, were now a tapestry of autumnal hues—russet, gold, and crimson—a fitting backdrop to the season of reflection that had settled over her. Once a source of overwhelming anxiety, the Grand Prix victory now felt like a distant echo, a testament to her remarkable journey. It wasn't the blue ribbon that held significance; it was the profound personal transformation that had accompanied it.

Looking back, she realized how deeply entrenched self-doubt had been. The relentless pressure to live up to her family's equestrian legacy, the constant comparisons to Emma's effortless grace and seemingly innate talent—these had weighed heavily on her. She'd allowed the fear of failure to paralyze her, to cloud her judgment, and to stifle her natural ability. The "center fence," that seemingly insurmountable obstacle, symbolized her internal struggles. Conquering it wasn't about clearing a jump; it had been about leaping over the invisible walls she'd built around herself.

The process of overcoming those internal barriers had been gradual and painstaking. It had required relentless self-reflection, a willingness to confront her vulnerabilities, and a conscious effort to reframe her perception of success and failure. She had learned to separate her worth from her performance, to appreciate the intrinsic value of the journey, regardless of the outcome. Wins and losses became less about proving herself and more about learning and growing. Each competition and training session allowed her to refine her skills, deepen her understanding of her horse, and strengthen her mental resilience.

Apollo, her magnificent chestnut stallion, had played a pivotal role in this transformation. He'd been more than a partner; he'd been a mirror reflecting her emotional state. His sensitivity, unwavering loyalty, and quiet strength had taught her invaluable lessons about patience, empathy, and resilience. Their bond had deepened through shared experiences, triumphs, and setbacks, forging a partnership built on mutual respect and unwavering trust. She understood now that true horsemanship wasn't simply about mastering technical skills; it was about cultivating a profound connection with the animal.

Isabelle, her seasoned trainer, had played a pivotal role in her development. More than a coach, Isabelle had been a mentor, a confidante, and a source of unwavering support. She'd seen beyond Leah's anxieties and recognized the exceptional rider hidden beneath the layers of self-doubt. Isabelle's guidance had extended beyond the riding arena, encompassing horsemanship, business acumen, and personal well-being. She'd instilled in Leah a profound appreciation for ethical practices within the equestrian world, emphasizing the importance of responsible horse care and the need for sustainable approaches.

Dr. Chen, the sports psychologist, had helped Leah unravel the complex web of her anxieties. Through cognitive-behavioral therapy and mindfulness techniques, she'd learned to manage the pressure of competition, control her racing thoughts, and maintain her focus under pressure. The tools she'd gained—breathing exercises, visualization techniques, and positive self-talk—had become essential to her mental training regimen. She'd discovered the profound power of the mind, its ability to shape her performance and overall well-being.

Dr. Ramirez, the veterinarian who focuses on equine health issues, significantly expanded Leah's understanding of equine well-being. Through studying equine anatomy, physiology, and biomechanics, she gained deep respect for the intricate needs of these amazing animals. Her commitment to equine welfare had blossomed into something more than a simple interest; The collaboration with Dr. Ramirez had helped to develop a comprehensive approach to horse care that prioritized the physical and mental health of her equine partners. The dream that she had harbored for so long, the dream of creating her own equestrian center, once an impossible aspiration, now felt remarkably close to becoming a tangible reality.

With painstaking effort, she had not only developed a comprehensive business plan and got the funding but also located the ideal site for her venture, a magnificent estate distinguished by its rolling hills and expansive grounds, perfectly suited to accommodate the extensive training facilities, paddocks, and stables required for her operation. She meticulously crafted the process, integrating advanced technology and sustainable approaches. Her horses' well-being was her priority, leading her to design spacious, naturally lit stables using eco-friendly materials and providing ample turnout areas. The project combined her passions–horses, riding, and leadership–into a unified purpose.

She envisioned a facility where new riders could master technical skills, learn about responsible horse care, and build strong character through ethical horsemanship. She aimed to foster a supportive community where riders of all skill levels could thrive, horse welfare was paramount, and collaboration was key. Leah found the community's strength during her journey. The intense and sometimes bitter rivalry that once existed between them had, over time, mellowed into a collaboration marked by mutual respect and understanding.

They had shared knowledge, offering each other support and encouragement. Leah had found a sense of camaraderie among her fellow riders, a network of support that transcended competition. She had mentored younger riders, sharing her hard-won lessons about resilience, self-belief, and perseverance. This shared experience enriched her understanding of the equestrian community, revealing the strength of collaboration and mutual respect.

The journey hadn't been without its challenges. There were moments of doubt when the sheer magnitude of her ambitions felt overwhelming. However, she'd learned to embrace these moments, seeing them as opportunities for growth. She'd developed coping mechanisms, drawing strength from her support network, her mentors, and her family. These moments of vulnerability had only strengthened her resolve, sharpened her focus, and reinforced her commitment to her goals.

As the final touches applied to her equestrian center, Leah felt a profound sense of gratitude. She'd overcome her self-doubt, harnessed her ambition, and transformed her passion into a tangible reality. The future remained unwritten, full of both challenges and opportunities. But Leah, armed with newfound self-belief, a strong support system, and an unwavering commitment to her values, was ready to embrace whatever lay ahead.

# Chapter 54: Continued Journey

The crisp morning air bit with a chill that hinted at the approaching winter, a stark contrast to the sun's warmth streaming through the large windows of her newly constructed equestrian center. Leah stood amidst the still-settling dust of the stables, the scent of fresh wood and hay mingling in the air. It was a tangible representation of her dreams, a testament to the unwavering dedication and sheer hard work that had brought her to this point. The polished wooden stalls, each meticulously designed for comfort and ease of access, stood in neat rows awaiting their equine occupants. The spacious paddocks with newly planted grass stretched beyond, promising ample space for the horses to graze and roam freely.

Hammering, sawing, and the rumble of machinery: The construction had been a whirlwind of activity, a symphony of sounds. There had been moments of doubt—the occasional setback, a missed deadline, the unexpected surge in material costs. But Leah, fueled by her vision and the unwavering support of her team, had navigated these challenges with remarkable resilience. Isabelle's mentor, who had been an invaluable source of guidance, offered her shrewd business advice and a calming presence during moments of stress. Together, they had meticulously managed the finances, ensuring the project remained on track despite unforeseen obstacles.

They scheduled the grand opening for the following month, and the excitement was palpable. They had sent invitations, secured sponsors, and the local media was excited. Leah had envisioned a center that went beyond the typical riding school, embracing the principles of responsible horsemanship, sustainability, and ethical practices. She'd incorporated eco-friendly materials wherever possible, using solar energy to power the facility and implementing water-saving irrigation systems for the paddocks. The design incorporated ample natural light, creating a bright environment for both horses and riders.

Beyond the physical structure, Leah equally focused on building a nurturing and supportive community. She had assembled a team of highly skilled instructors, each possessing a passion for horses and a commitment to ethical training methods. She had also contacted local organizations, fostering partnerships to provide educational programs for children and adults and promote responsible horse care and environmental stewardship to build a center for equestrian enthusiasts of all ages and skill levels to learn, grow, and connect.

The selection of her first horses had been a meticulous process. She had sought horses with exceptional temperaments, prioritizing their physical and mental well-being above all else. She had chosen a diverse group, including experienced show jumpers, talented dressage horses, and gentle riding ponies suitable for young children. They carefully assessed each horse, not for its athletic prowess, but for its compatibility with the training program and its overall personality. The goal was to create a harmonious herd in which each horse felt safe, respected, and cherished.

Apollo, of course, would be the central figure. He was more than a horse; he was Leah's trusted companion and a symbol of her journey. His calm demeanor and unwavering loyalty made him an ideal role model for the younger riders. Leah planned to use him in demonstration lessons, showcasing the principles of classical dressage and showing the beauty and artistry of the discipline. His presence would be a constant reminder of the importance of nurturing a deep bond with one's equine partner.

The memories of the Grand Prix still resonated deeply within her. It wasn't the victory but the transformation she had undergone. The journey had taught her the invaluable lessons of perseverance, self-belief, and the power of the community. She had learned to manage her anxieties, to embrace her vulnerabilities, and to find strength in her relationships with others. The competitive spirit remained, but it was now tempered with a profound appreciation for the ethical considerations within the equestrian world.

Emma, once her fiercest rival, had become a valued friend and colleague. Their initial rivalry, born out of ambition and a shared desire for success, had developed into a mutual respect, a partnership built on shared goals and a common passion for horses. They regularly exchanged ideas, offering each other support and guidance. Leah valued Emma's insightful perspective and

Appreciated her willingness to collaborate on projects benefiting the equestrian community. They planned to organize joint clinics and competitions, fostering a spirit of camaraderie and teamwork among their students.

Leah's relationship with Isabelle continued to flourish. A seasoned equestrian, Isabelle offered steadfast support and guidance, especially during the challenging early days of establishing the center. Her mentorship went beyond business, influencing every aspect of Leah's personal and professional growth. Isabelle's wisdom and experience helped shape Leah's approach to leadership. The support from her family remained an unwavering pillar. Her parents, initially hesitant about her ambitious plans, had become staunch supporters, offering both emotional encouragement and practical help. They understood the immense personal growth she had undergone and acknowledged the profound impact the equestrian world had had on her life. Their pride in her achievements was clear, providing a constant source of reassurance and motivation.

Dr. Chen, the sports psychologist, continued providing guidance, helping Leah manage the stresses and pressures of running the center. The mindfulness techniques she had learned proved invaluable in navigating the challenges of entrepreneurship. Maintaining focus, controlling her emotions, and remaining grounded in her values was crucial in creating a successful and ethical business. Dr. Chen's support was a constant reminder of the importance of mental well-being in all aspects of life.

Dr. Ramirez, the equine veterinarian, remained a valuable resource, offering expert advice and guidance on all matters concerning the horses' health and well-being. Leah's commitment to responsible horse care ensured every horse under her care received the highest quality veterinary attention. She understood the delicate balance between athletic performance and equine health, working closely with Dr. Ramirez to develop a comprehensive wellness program. This program prioritized preventative care and early intervention, minimizing the risk of injury and maximizing the well-being of each horse.

Looking ahead, Leah felt a profound sense of accomplishment. The equestrian center was not a building, but a testament to her resilience, unwavering commitment, and belief in the transformative power of horses. It was a place where young riders could learn to ride and cultivate a deeper understanding of these magnificent animals, as well as the ethical responsibilities that accompany equestrian pursuits. The journey had been challenging, filled with moments of doubt and setbacks. But through it all, she had learned to embrace the challenges, to find strength in her vulnerabilities, and to value the importance of community and collaboration. She understood that the path ahead was certain to bring about a multitude of new adventures, a series of increasingly challenging situations that would demand her best efforts, and an abundance of opportunities for personal and professional growth. A newfound sense of purpose and fulfillment filled her, and with Apollo at her side, she felt prepared and excited to ride bravely into the unknown.

# Chapter 55: Epilogue: A Glimpse Ahead

The scent of earth and pine clung to Leah's clothes, a reminder of the frantic final days of construction. Evenings had been spent on blueprints, invoices, and calming her team. Now, at the threshold of her dream, a quiet contentment settled over her. The grand opening captured the whirlwind of emotions from the past year. In the staff kitchen, only the hum of the refrigerator competed with Apollo's steady hoofbeats in the nearby paddock.

As the sun cast a warm glow on the outdoor arena in the late afternoon, she ran her fingers along the smooth grain of the newly polished wooden railing, finding solace in the cool contrast of the wood against the sun's heat. Ready and waiting for the event, stood the arena; a testament to meticulous planning and a true masterpiece of engineering and design. The sand, raked to a flawless and pristine condition, gleamed brilliantly under the sunlight, patiently awaiting the first hoofbeats of the horses and their riders, who would be the first to leave their mark upon its smooth surface. The jumps, meticulously placed as a testament to both Emma's keen eye for detail and their shared artistic vision, stood sentinel-like, perfectly poised and ready for the athletes to begin their routines.

Emma, an invaluable partner throughout the design and construction phases, leveraged her extensive experience as a top competitor to achieve the ideal fusion of functionality and aesthetic appeal in the project. Their collaboration, initially a business venture, had blossomed into something far exceeding mere professional dealings.

However, the future didn't solely comprise sun-drenched arenas and the quiet satisfaction of a job well done. The business side held its anxieties. While the initial response had been overwhelmingly positive, the reality of maintaining a thriving equestrian center proved to be a formidable undertaking. Securing consistent funding, managing the diverse needs of staff and clients, and ensuring the health and well-being of the horses were all significant, ongoing concerns. The pressure, though different from the intense focus of competition, was no less demanding.

Dr. Chen's words echoed in her mind—the importance of mindfulness, of recognizing the subtle shifts in her emotional landscape. The techniques she had learned hadn't merely helped her conquer her pre-competition nerves; they were now essential tools in navigating the complexities of entrepreneurship. She took quiet moments throughout the day, breathing deeply, centering herself amidst the chaos. It was a conscious choice, a dedication to self-care that felt as essential as the meticulous care she provided for her horses.

A familiar tension tightened in her shoulders as she sat reviewing financial reports one evening. The weight of responsibility and the fear of failure crept in, a persistent undercurrent beneath the surface of her success. She knew, however, that she couldn't let these anxieties paralyze her. She reached for a small, smooth stone that lived in the pocket of her riding breeches—a memento from her grandmother, a symbol of resilience. Holding the cool stone in her hand, she focused on her breath, letting the tension slowly seep away.

However, doubt spread beyond the business world. A new challenge loomed that had nothing to do with budgets or marketing. A local landowner, Mr. Fitzwilliam, known for his stubbornness and a deep-seated resistance to change, threatened to build a large-scale development directly next to her property. The development, which included a highway and several housing complexes, would dramatically alter the landscape, create noise, and light pollution. This would severely affect the tranquil environment, which is crucial to the horses' well-being and the success of her equestrian center.

The news had struck Leah like a blow. She knew that fighting Mr. Fitzwilliam would be a formidable battle. He was an influential figure in the community, well-connected and fiercely protective of his interests. Yet, the thought of compromising her vision, of jeopardizing the sanctuary she had created, was unacceptable. This was more than a business; it was a dream, a testament to her passion and dedication. She had poured her heart and soul into cultivating a harmonious environment for horses and humans.

The subsequent legal battle proved to be an arduous ordeal, marked by long days filled with meticulously reviewing countless documents, collaborating intensely with her lawyer to develop effective strategies, and skillfully navigating the complex maze of zoning laws and environmental regulations, testing the limits of her resilience in ways that she had never previously conceived. Emma, Isabelle, and her entire family provided support that proved to be invaluable to the success of the project. They rallied around her, offering not only words of encouragement to bolster her spirits, but also providing the practical help she desperately needed to navigate this challenging time. Because they believed in her strength and offered unwavering support, she found the fortitude to persevere through challenges.

Simultaneously, she faced a new challenge in the riding arena. A young, prodigiously talented rider named Alex emerged on the competitive scene, her skill and determination matched only by her fiery spirit and unwavering ambition. Alex presented a unique challenge to Emma; this wasn't a rivalry born of personal friction, but a clash of ambition and talent. Alex's presence sparked a new fire within Leah, a renewed desire to push her boundaries and show her mastery of the sport. This competition was less about proving herself to others and more about continually developing and improving her ability.

Amid the legal battles and the competitive arena, a new chapter in her personal life unfolded unexpectedly. A kind and understanding veterinarian, Liam, who often worked with Dr. Ramirez, became a close friend. His calm demeanor and shared passion for horses forged a bond beyond professional respect. Their conversations grew to include shared dinners, quiet walks along the center's trails, and a deeper connection. This new relationship brought balance to her life, a reminder that joy and love could be steady sources of strength amid challenges.

# Acknowledgments

First and foremost, I'm deeply grateful to all the horse lovers who inspired this story. Your passion and dedication to these magnificent animals constantly fill me with wonder and admiration. The Youth Cutting Horse members, a talented group of young artists, deserve special recognition for their vibrant paintings that brought Star and Alex's story to life. The narrative was uniquely and captivatingly shaped by her creative vision, adding an extra dimension to the tale.

To my family, whose unwavering support and encouragement made this book possible. Thank you for believing in my dreams and sharing in the journey. Your love and faith have been my guiding light, and I am forever grateful for your presence. You have been my rock, my cheerleader, and my inspiration. This book is as much yours as it is mine.

I'd also like to thank my illustrator, Karen Shayler, and my editor, Roxana Coumans, for their indispensable help and support during the book's publication. This story wouldn't be alive without your keen eye, insightful feedback, and endless patience. This project has benefited from your dedication to excellence and belief in its success.

I'm also thankful for the invaluable guidance and expertise provided by my reviewers, including my mother, Billie Aylesworth of Diamond B Cutting Horses. Your feedback and insights have added authenticity and depth to the story. I'm constantly inspired by your love of horses and commitment to excellence.

To all the readers who have embarked on this journey with Alex and Star, thank you for your support and enthusiasm. Your love for stories and your belief in the magic of friendship and perseverance have made this book a reality. May this tale inspire you to chase your dreams, nurture your friendships, and cherish the bonds that enrich your life.

# About the Author

Brett Shayler, an author, has been passionate about horses and the natural world his entire life. Raised on a ranch surrounded by diverse animals, his storytelling vividly reflects the profound and enduring connections between people and the animals they cherish and care for throughout their lives. Drawing inspiration from his family's vast acreage and the wild horses that roam freely upon it, Brett lovingly crafts heartwarming tales that not only celebrate the bonds of friendship but also extol the virtues of a deep respect for the natural world and the remarkable connections that exist in the relationships between people and animals. Through his books, he hopes to cultivate curiosity, compassion, and a thirst for knowledge in young readers. Brett often escapes to nature, hiking and observing wildlife, finding inspiration for his writing away from his desk. Brett proudly holds membership in both the prestigious National Cutting Horse Association and the equally esteemed American Quarter Horse Association, demonstrating his deep commitment to these organizations and the equestrian world.